Hairdressing Theory

A Simplified Approach

R J Jeynes

Lecturer in Hairdressing, West Notts. College

Stanley Thornes (Publishers) Ltd

First published 1984 by
Stanley Thornes (Publishers) Ltd,
Educa House,
Old Station Drive,
Leckhampton,
CHELTENHAM GL53 0DN

British Library Cataloguing in Publication Data

Jeynes, R.
    Hairdressing theory.
    1. Hairdressing
    I. Title
    646.7'242   TT958
    ISBN 0-85950-140-X

Typeset by Artmark, Nailsworth, Gloucestershire.
Printed and bound in Great Britain by The Pitman Press, Bath.

Contents

permanent waving method. Machine perming. Style supports. Recognizing faults in permanent waving. Testing for compatibility.

Preface

In a world where machines, computers and robots are increasingly replacing traditional craft skills, few occupations can offer the creative satisfaction enjoyed by those employed in the craft of hairdressing. The skilled hairdresser requires more than artistic ability — he must be an expert technician, a salesman, a consultant and a communicator.

Enthusiasm is the most important attribute of any hairdresser, whether he be a student embarking on a career in hairdressing, or a qualified stylist aiming to maintain and increase his skills in a highly competitive and ever changing profession.

The wide variety of services offered in a modern salon require considerable practical expertise, much of which can be, and is, best learnt by seeing and doing. This practical ability will, however, be greatly limited if the relevant theory is not also learnt.

This book, as its title suggests, aims to provide, in an easy to understand way, the knowledge necessary to approach confidently every aspect of salon routine.

R J Jeynes

1

Understanding hair

WHAT IS HAIR?

Hair is a modified extension of the skin. It is a dead material, containing neither blood vessels nor nerves, although it is part of the nervous system. Composed of 45% carbon, 30% oxygen, and lesser quantities of hydrogen, nitrogen and sulphur, hair is similar in property to fingernails and toenails. It is a very strong and resilient material. To give some indication of this strength, one human hair is capable of suspending a weight of up to 113 g (4 oz). This is about five times the weight which a piece of copper wire the same thickness would hold. When wet, it can stretch by 30% or more, and it is this stretching action which helps us to set or blow-dry hair into different styles.

HAIR GROWTH

At the base of each hair *follicle* (a small cavity in the skin through which the hair grows) is a living hair bulb. This bulb is composed of cells which rapidly divide to produce the components of hair fibre. These cells are forced up the follicle, continually changing shape, losing moisture and becoming hard. They are then joined together by a system of cross-linkages known as disulphide bonds. *Keratinization* is the name given to this hardening process.

Hair usually grows at the rate of 12 mm ($\frac{1}{2}$ in) per month, although this slows down as one becomes older and also during the colder, winter months. According to its texture, it has a thickness of between 0.02 to 0.05 mm (1/15 000 to 1/500 in). The average area of a scalp is 775 cm² (120 in²), with approximately 155 hairs per cm² (1000/in²), but this number varies according to the texture and colour of the hair. The darker the colour, the stronger the texture. A head of very dark hair will be covered by approximately 100 000 hairs whereas a head of very blonde hair will be covered by up to 140 000 hairs. On average, a single hair will grow for 3 to 4 years before being released from the follicle (although in many cases it can continue growing for up to 6 years or more). Once released, it is replaced by a new hair which has already begun to form in the follicle.

THE STAGES OF HAIR GROWTH

There are three stages of hair growth.

1 *Anagen phase*
More than 80% of all hairs on a healthy scalp are in the anagen phase at any given time, this being the most active stage of hair growth. When observed under a microscope the hair bulb appears dark in colour (unless the hair growing is white). This is due to the production of *melanin* which is the hair's colour pigment. The anagen phase continues for 2 to 6 years.

2 *Catagen phase*
This is a transitional stage when the activity of the bulb slows down, producing fewer new cells until it ceases completely. About 1 to 2% of all hairs on the scalp are in the catagen phase at any given time.

3 *Telogen phase*
The bulb has a resting period during this final stage and the hair is released from the follicle. About 75 to 125 of these hairs collect in your brush every day.

Hair cells are made up of proteins which are collectively known as *keratin,* with the addition of small amounts of carbohydrates, oils, minerals and lipids. Hair is *hygroscopic* (in other words, it absorbs moisture from the atmosphere) and, depending on the conditions of humidity, the hair can hold 15% or more of its total weight in water.

THE THREE LAYERS OF HAIR

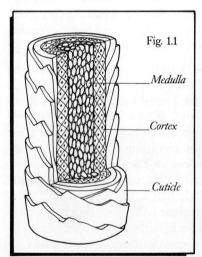

Fig. 1.1

Medulla

Cortex

Cuticle

The outer layer of the hair shaft is called the *cuticle.* It has a hard, scale-like structure, with anything up to 11 layers. The cuticle creates a protective shield for the more delicate internal structure. When in good condition, it reflects light and gives hair a shiny appearance. One of the effects of ammonia in perming and tinting preparations, is to break down this barrier, open the cuticle layers, and allow penetration to the deeper layers. It is vitally important that the cuticle layers are closed again following such treatments (See p. 34).

The central section of the hair is the *cortex* and it is here that the colour pigment is contained. The cortex is made up of a fibrous substance, of elongated, proteinous cells, and in the living section of the hair the cortex contains fluid.

The *medulla* is the core which runs through the middle of the hair shaft, but it is not found in all hair. It appears to have very little effect on the hair's strength and its function is still to be clearly determined.

EFFECT OF DIET ON HAIR GROWTH

There is a saying that 'We are what we eat', and that is certainly true. Protein is necessary for the healthy growth of our bodies and also for healthy hair growth; in fact good hair condition starts with a well-balanced diet. Foods which are particularly good in this respect are fish, cheese, eggs and meat, and a diet lacking such food-stuffs will not only affect the hair's condition, but the growth rate will also be slowed down.

HAIR SHAPE

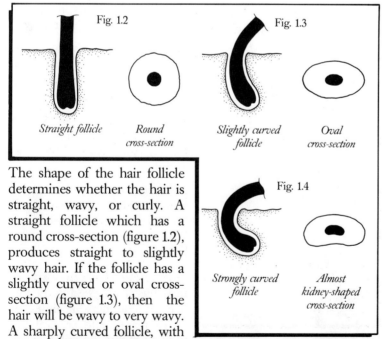

Fig. 1.2

Straight follicle

Round cross-section

Fig. 1.3

Slightly curved follicle

Oval cross-section

Fig. 1.4

Strongly curved follicle

Almost kidney-shaped cross-section

The shape of the hair follicle determines whether the hair is straight, wavy, or curly. A straight follicle which has a round cross-section (figure 1.2), produces straight to slightly wavy hair. If the follicle has a slightly curved or oval cross-section (figure 1.3), then the hair will be wavy to very wavy. A sharply curved follicle, with an almost kidney-shaped cross-section (figure 1.4), will produce curly to very curly hair.

TECHNICAL TERMS GIVEN TO HAIR

Hair falls into two main categories: *vellus* is the name given to the soft hair which covers most areas of the body, and *terminal* is the coarse hair found on the head, armpits and pubic areas, and in the case of a man, forms the beard, moustache and hairs on the chest.

Terminal hairs can be sub-divided as follows:

Barba   — the face

Capillus   — the scalp

Cilia   — the eyelashes

Supercilia   — the eyebrows

Vibrissae   — the nostrils

Tragi   — the ears

Hirci   — the armpits

Pubes   — the pubic region.

Hair is found on all parts of the body except the palms of the hands, the soles of the feet, and the lips.

Scalp and Hair Diagram—figure 1.5

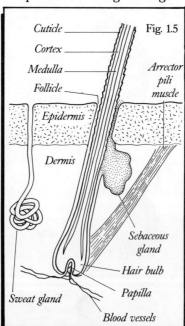

Follicle A small cavity in the skin through which the hair grows.

Epidermis The skin's outer protective layer of dead skin cells, which are continually being shed and replaced.

Dermis The layers of living skin cells which are continually replacing the cells of the epidermis.

Sebaceous Gland The gland which produces the skin's natural oil called sebum. This is fed into the hair follicle and then on to the scalp.

Arrector Pili Muscle An involuntary muscle attached to the follicle. It contracts at times of shock, fear or cold, erecting the follicle and causing a roughened condition of the skin called goose flesh. This action reduces heat loss from the body.

Hair Bulb The part of the hair which fits over the papilla. It is composed of soft, growing cells.

Papilla This contains the constituents necessary for the production of hair.

Sweat Gland A gland which produces a mild, saline solution, which cools the body by evaporating from the surface of the skin. It also acts as an excretory organ to assist in the removal of harmful waste products from the body.

Blood Vessels These provide nutrition necessary to sustain or grow a hair. They also carry hormones which govern the growth and lifecycle of a hair.

2

Hair and scalp disorders and diseases

It is important that the hairdresser is able to recognise any problems with the hair or scalp, and is able to offer suitable advice to the client. Certain conditions, such as oily or dry hair, or hair damaged by chemicals, can be treated in the salon, but wherever infestation of the scalp is suspected, eruptions of the skin or inflammation are present, the client *must* be referred to a doctor.

A protective acid mantle covers the entire body, preventing the growth of bacteria which generally require an alkaline state in which to flourish. A proportion of this acid mantle is made up of the skin's natural oil – *sebum* – and if for some reason this is weakened or disturbed, there is a distinct possibility that bacterial infection will occur. Most of the problems encountered by the hairdresser are caused by an over or under-production of sebum and there are many contributory factors which can bring about this imbalance, the most obvious being the client's general state of health. The hair and scalp act as a very good barometer when it comes to forecasting how healthy or unhealthy we are. They are affected by a wide variety of causes, from a simple cold to a poor diet. Keeping active, having plenty of fresh air and eating wisely, are all of tremendous value in helping to maintain a healthy hair and scalp condition.

COMMON PROBLEMS

Hair loss (Alopecia) The term alopecia is applied to various forms of hair loss, which may be temporary or permanent.

Alopecia Senilis This is the term for male baldness due to old age.

Alopecia Areata This term is used to describe a condition where a group of hairs suddenly fall out, leaving round or oval-shaped patches of baldness. These patches can appear in a matter of hours and may grow to leave quite large areas of bare skin. This condition is usually attributed to a nervous disorder and in the majority of cases it will grow again although it may be white.

Alopecia Totalis This is the loss of hair from the scalp and the rest of the body.

Traction Alopecia This refers to patches of baldness usually seen around the hairline and at the crown, caused by rollers which have been wound too tightly and which have pulled the roots.

Alopecia Universalis This term describes the *total* loss of hair from the entire body.

Alopecia Prematura This is a form of premature baldness which is generally confined to men and is often hereditary. It occurs around the frontal area of the head and also at the crown.

Trichotillomania This is another form of hair loss, but one which is self-induced as a result of nervous tension. It is a habit commonly found among children. They pull out hairs from their head and are left with irregular bare patches.

Various types of drug can also cause hair loss which tends to be from all over the head, as opposed to a total loss from one or more small areas. Hair breakage can be distinguished from hair loss, by examining the hairs collected in a brush. If no white bulb is visible at the root end, the indication is that the hair has broken off, and a close examination of the scalp will often show areas of stubble. The main causes of this form of hair loss are due to mechanical or chemical abuse, such as the effects of an incorrect permanent waving technique or harsh backcombing.

Pityriasis Simplex This term is generally referred to as scurf or dry dandruff. The condition can be recognised by flakes of skin breaking away from the surface of the scalp, which when the hair is brushed, fall on to the shoulders. It is not a serious condition unless the scaliness is very marked, but from a cosmetic point of view it can spoil one's appearance.

Pityriasis Steatoids This is a more advanced form of pityriasis simplex, where the scales tend to stick to the scalp with grease, and are yellowish in colour and crusty to the touch. If the scales are removed from the scalp, the skin underneath may appear pink. No irritating agents should be used on a scalp suffering from this condition.

Seborrhoea Oleosa This term describes acute dandruff brought about by an excessive discharge of sebum. The hair becomes very lank and greasy and the scalp may be covered by thick sticky scales which are an ideal breeding ground for bacteria. The face may also become markedly greasy, accompanied by acne, and owing to the fact that a follicle filled with sebum will tend to release its hair more quickly during the final stage of a growth cycle, abnormal hair fall may be noticed.

Fragilitas Crinium The splitting of the hair shaft which is normally seen at the ends of the hair. It can occur in greasy as well as dry hair.

Canities This is brought about by an alteration in the balance of the colour pigment, melanin, and the air spaces in the hair shaft, causing the hair to turn white.

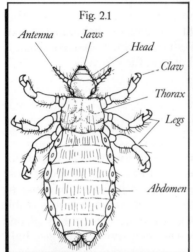

Fig. 2.1

Antenna Jaws
Head
Claw
Thorax
Legs
Abdomen

Pediculosis Capitis A disease of the scalp caused by an infestation of small insects known as pediculus capitis (head lice). (See figure 2.1) Fully grown head lice are about 2 mm ($\frac{1}{12}$ in.) long and are parasites, feeding off blood which they suck from the scalp. They have a life span of around 30 days during which time the female lays her eggs close to the scalp; these eggs are recognisable as white, oval-shaped specks, 1 mm in length – they are referred to as nits.

Contrary to popular belief, head lice do not confine themselves to living on a dirty scalp but are equally at home on the cleanest of heads. Neither do they jump through the air from one head to another, although they can be transferred from one person to another by direct contact or by intermediate means, such as a towel or comb which has been in contact with an infested person. This problem must be medically treated and under no circumstances should the hairdresser continue working with an infected head. All items which have come into contact with the client, such as gowns, towels, brushes, combs, etc., must be thoroughly sterilized.

Impetigo This is a contagious bacterial infection which again must be medically treated. Impetigo is usually confined to children and is found on the face as well as the scalp. It is caused by bacteria penetrating the skin, following which blisters filled with fluid develop. The fluid dries and thickens producing honey-coloured crusts.

Ringworm This is a serious fungal infection which is normally confined to children and requires medical treatment. It first appears as circular patches of very short grey scales. It is extremely contagious.

Psoriasis The symptoms of this condition are dry, hard, silvery scales with a reddish border. It is not confined to the scalp but may also be found on the knees and elbows. The cause of psoriasis is speculative, one theory being that it is brought about by imbalances in a person's metabolism, and no known cure is available. It is neither infectious nor contagious.

Dermatitis An acute inflammation of the skin which can appear at any time. It is caused by an irritant coming into contact with the skin and is a condition to which the hairdresser should pay very special attention, not only from the point of view of the clients' safety, but his own safety too. The symptoms are a redness of the skin which may be followed by blisters and running sores. It will occur where a person is allergic to a particular substance or chemical, which can be anything from a bar of soap to permanent tinting preparations. Although the condition often clears up in a few days, medical advice should be sought to establish the cause. A chronic condition may arise where the irritant has been continually applied or in unusual cases after only one application. At its worst, it can cause fever, vomiting and temporary blindness. Because of the risks involved, it is important that certain precautions are taken to ensure that both client and operator are protected from contracting dermatitis. Wherever the manufacturer of a product states the necessity for a skin test to be carried out prior to the product being used on a head, *this must be done.* The most common cause of dermatitis in the salon is brought about by the para-dyes contained in permanent tinting preparations. The operator is best protected by wearing rubber gloves whenever irritation may be expected such as for permanent waving and colouring.

3

Hairdressing-related chemistry

ELEMENTS

The simplest form of matter is referred to as an *element*, and is made up from particles *(atoms)* which are all the same; carbon, sulphur and iron are three examples. However much an element is broken down, it will remain the same substance until the atoms are split in some sort of nuclear reaction. Everything around us is made from elements (there are approximately 100 of them), and these fall into two main categories, metallic and non-metallic elements.

Some common examples of elements

Metallic	Non-Metallic
Magnesium	Oxygen
Sodium	Carbon
Calcium	Nitrogen
Zinc	Hydrogen
Copper	Sulphur
Potassium	Chlorine
Gold	
Iron	
Nickel	

COMPOUNDS

There are two forms of compound: *organic* and *inorganic*. Organic compounds are the basis of living organisms and they all contain carbon; inorganic compounds constitute all other compounds. Compounds are composed of elements and in the case of organic compounds, are often of a very complex structure. Inorganic compounds on the other hand, are generally very simple substances containing no more than three different elements.

Some common examples of compounds

Organic	Inorganic
Sugar	Sulphuric acid
Soap	Salt
Alcohol	Water
Keratin	
Proteins	
Melanin	

SALTS

A salt is produced by combining an acid with an alkalki. Ordinary domestic salt is sodium chloride – a combination of hydrochloric acid and sodium hydroxide.

Salts are usually crystalline, they have a neutral pH of 7 (see p.16).

Ammonium thioglycollate is a salt, which is used in the manufacture of permanent wave lotions. It is formed by the combination of thioglycollic acid and ammonium hydroxide.

Some other examples of salts

Salt	Examples of usage
Calcium carbonate	– Toothpastes
Potassium carbonate	– Soaps
Potassium permanganate	– Antiseptics
Borax	– Water softeners and baby soaps
Sodium carbonate	– Bath salts and water softeners
Zinc oxide	– Calamine lotion

EMULSIONS

When we have two substances which will not dissolve into each other, such as oil and water, it is possible to use an *emulsifying agent* which mixes them together so that they form a milky or creamy substance. Emulsifying agents are usually either water-soluble or oil-soluble gums, resins or sulphonated fats. Nearly all creams, such as hair cream, cold cream, cream tints, etc., are emulsions. An oil-in-water emulsion (O/W) encourages oil globules to disperse in water. A water-in oil emulsion (W/O) encourages water globules to disperse in oil.

MIXTURES

Unlike a compound, no chemical change takes place in the formation of a mixture and its constituents can be separated by simple physical processes.

Mixtures have no fixed composition; they can contain two or more substances, which can be either compounds or elements.

OILS

Vegetable oils are extracted from plant sources and are used in many hairdressing preparations and toilet soaps. They are insoluble in water and alcohol (with the exception of castor oil which is soluble in alcohol) and are edible. Shampoos for dry hair often contain almond oil, coconut oil, or olive oil.

Essential oils are part of the vegetable oil family. They are essences, extracted from plants, flowers, leaves, seeds, and the bark of trees, and are widely used in the manufacture of perfumes.

Mineral oils, unlike vegetable oils, are not digestible. They are extracted from the ground and their big advantage over vegetable oils when used in hairdressing preparations is that they will not go rancid. They are insoluble in water and most are insoluble in alcohol, and are used to a very large extent in the manufacture of polishes, lubricants, and dressings such as brilliantine.

SOME PRINCIPAL CHEMICALS USED IN HAIRDRESSING

Sulphur One of the most important and widely use chemicals in the world today. Sulphur used to be known as 'brimstone' and has been used as an antiseptic for a considerable period of time. It forms the base of many skin oils and lotions, and is also used in anti-scurf lotions and preparations for treating greasy scalp conditions. In its natural state, sulphur is a light-yellowish powder, which is insoluble in water.

Salicylic acid An antiseptic and preservative, which is frequently used in the treatment of skin diseases.

Citric acid This is obtained from many varieties of fruit and is a white, crystalline substance. Citric acid is often used in cream rinses and products designed to restore the hair's natural acid balance (its correct pH, see p. 16) following chemical treatments.

Amyl acetate A solvent used in the manufacture of lacquers and varnishes, such as nail varnish, and also as a varnish remover.

Acetone This is another solvent for resins and varnishes.

Sulphonated oil This is obtained when an oil has been made to react chemically with concentrated sulphuric acid. One of the most popular sulphonated oils is one combined with castor oil. They are used in the manufacture of soapless shampoos and permanent wave lotions, the advantage being that they are soluble in water.

Alcohol Pure alcohol is used in perfumes and in the manufacture of setting lotions.

Ammonia liquid This is a solution of 33% ammonia gas in water. It is used in permanent waving lotions and tinting products. Any product which contains ammonia should always be mixed in a non-metallic bowl, to prevent a pungent, choking gas being created.

Distilled water This is ordinary water purified by distillation. It is used in the preparation of most solutions requiring water and should always be used when diluting hydrogen peroxide.

Iodine This product has many sources, one being from the ash of seaweed. Tincture of iodine is used as an antiseptic on cuts and wounds, and for the treatment of certain scalp infections.

Henna powder Made from the powdered dry leaves of the Egyptian privet, it is one of the oldest forms of hair colourant. Compound hennas are a manufactured combination of pure henna plus other vegetable or metallic dyes.

Shellac The refined secretion from the lac insect, used in the manufacture of hair lacquers and varnishes.

SYMBOLS FOR SOME COMMONLY USED CHEMICALS

O	— Oxygen	Al	— Aluminium
H	— Hydrogen	Ca	— Calcium
C	— Carbon	Fe	— Iron
N	— Nitrogen	Cl	— Chlorine
K	— Potassium	Na	— Sodium
S	— Sulphur	Mg	— Magnesium

HYDROGEN PEROXIDE

Hydrogen peroxide is one of the most commonly used chemicals in the salon. It has a similar chemical symbol to water, H_2O, but with an extra atom of oxygen H_2O_2. Tints and bleaching products contain an alkali, which disengages oxygen atoms from the peroxide to bring about a more rapid oxidation process. As well as being in liquid form, cream peroxide is also available. It is more stable than liquid peroxide but cannot be used as part of a permanent wave neutralizer.

Volume

The strength of peroxide is indicated by the term *volume*. The most common strengths used in the salon are 10, 20 and 30 volume. This tells us, for example, that 1 cc of 10 volume peroxide releases 10 cc of oxygen. By the same token, 1 cc of 20 volume peroxide releases 20 cc of oxygen and 1 cc of 30 volume peroxide releases 30 cc of oxygen.

If we pour out 40 cc of 20 volume peroxide, the amount of available oxygen would be 800 cc (amount of peroxide, 40 cc x peroxide strength, 20 volume). However, when peroxide is mixed with a tint, it becomes 'diluted' and the amount of available oxygen is reduced.

Example A bottle of tint, containing 40 cc, is mixed with 40 cc of 20 volume peroxide. The total mixture is therefore 80 cc. We know that 40 cc of 20 volume peroxide releases 800 cc of oxygen (40 x 20), but to find out how much oxygen is available now that it is mixed with the tint, we divide the 800 cc by the quantity of the total mixture (80 cc):

$$\frac{800}{80} = 10 \text{ volumes of available oxygen}$$

PEROXIDE DILUTION CHART

VOLUME STRENGTH	STRENGTH REQUIRED	PARTS PEROXIDE	PARTS DISTILLED WATER
60 vol	30 vol	1	1
40 vol	30 vol	3	1
40 vol	20 vol	1	1
40 vol	10 vol	1	3
30 vol	20 vol	2	1
30 vol	10 vol	1	2
20 vol	10 vol	1	1

TESTING PEROXIDE STRENGTH

Testing the strength of liquid hydrogen peroxide is done by using a peroxometer (figure 3.1). This is a glass float, calibrated from 0 to 100 volume, which is supplied with a glass measuring cylinder into which the peroxide to be tested is poured.

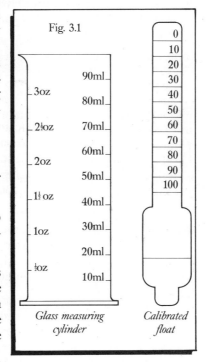

Fig. 3.1

Glass measuring
cylinder

Calibrated
float

Method

1 Fill the special glass cylinder to 85 ml (3 fluid oz).

2 Place the peroxometer into the liquid and spin it to disperse any air bubbles.

3 When the peroxometer has stopped spinning, take the reading from the line which is level with the surface of the liquid. This is the peroxide strength in volume.

4 On completion, rinse the peroxometer, dry thoroughly, and return it to the protective case.

Some Points to Remember

1 Always store peroxide in a cool, dark place.

2 Never leave the cap off the bottle longer than is necessary.

3 Never re-bottle peroxide which has been poured out for use.

4 Avoid contact with skin and eyes.

5 Always use the strength of peroxide recommended by the manufacturer.

THE pH SCALE

pH is a chemical measure which means power of hydrogen. This is the hydrogen ion content of a solution. By using a pH scale, we are able to determine the strengths of acids and alkalis.

Weak acids such as fruit juices, are beneficial to the human body, yet very strong acids, such as sulphuric acid, are extremely dangerous. Similarly, weak alkalis can be of benefit to us (for example, the use of bicarbonate of soda for the relief of an upset stomach), whereas a strong alkali such as caustic soda (used for oven cleaning) is highly poisonous.

Although there are a number of ways of testing the pH of a solution, the most accurate method is to use a special electronic meter.

The meter is numbered from 0 to 14, number 7 representing neutral (based on pure water), and it is on this number that the needle of the meter rests. When a solution is being tested, the needle will move to a point lower or higher than 7, indicating whether the solution is acid or alkaline. Any reading between 0 and 7 shows that the solution is acid (the lower the number, the stronger the acid), and any reading between 7 and 14 shows it to be alkaline (the higher the number, the stronger the alkali).

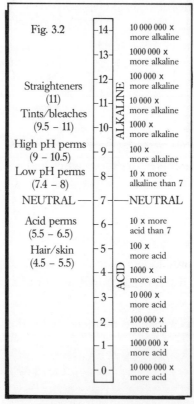

Fig. 3.2

Universal test papers are an easy way to check the strength of solutions in the salon. When dipped into a solution, the papers change colour and by comparing this to the colour code provided, the approximate pH can be established.

As can be seen from figure 3.2, hair is acidic, having a pH value of between 4.5 and 5.5. It is important that this acidic state is maintained, particularly following any chemical treatment. Leaving the hair in an alkaline condition is not only harmful to it but is also detrimental to the end result of the process carried out. Products known as antioxidants are manufactured for use following a bleach or tint application. They return the hair to its correct acid state, leaving it shiny and conditioned, and help to reduce colour fading.

Poor-quality conditioners, hairsprays, setting lotions and shampoos can be in an alkaline pH range, and with constant use will be detrimental to the condition of the hair. They should be avoided.

Temperature has an effect on pH. It is raised by higher temperatures and lowered as conditions become cooler.

4

Electricity

Electricity is a form of energy which can be converted into heat, light and power, for operating many different appliances.

The role played by electricity in the salon is a vital one, and a basic knowledge of it is necessary, even if for no other reason than being able to change a plug, or more seriously, knowing what course of action to take in the event of someone receiving an electric shock.

Electricity for domestic and most industrial uses is produced in a power station. It is carried to wherever it is needed by cables suspended from pylons high above the ground, or through cables buried underground (figure 4.1).

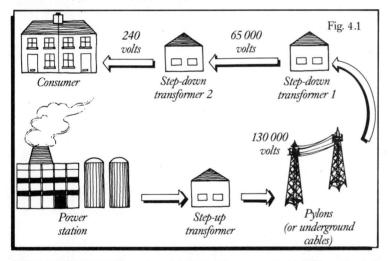

Fig. 4.1

For maximum efficiency, the electricity has to be transmitted at a very high voltage. Before it reaches the consumer, it passes through a series of step-down transformers to reduce the voltage.

The mains voltage throughout the United Kingdom is 240 volts, but in many other countries it is only 110 volts. Some equipment is designed to be used on either voltage, a converter switch being incorporated in the outer casing. Under no circumstances should an appliance designed for use only on a 110 volt supply be used on a 240 volt supply.

Voltage may be thought of as 'electrical pressure'.

The 'flow' of electricity is called the *current*. There are two sorts of electric current — alternating current and direct current.

ALTERNATING CURRENT (AC)

Salon and domestic appliances work on this sort of current. Alternating current, as its name implies, flows in both directions alternately. Mains electricity alternates at a frequency of 50 cycles per second; this is shown on a plate fixed to the appliance as 50~.

DIRECT CURRENT (DC)

Direct current, on the other hand, only flows in one direction. Examples of direct current are the current supplied by a battery, and the current required to work the amplifiers and picture tube in a TV set (the mains alternating current needs to be changed into direct current for this purpose). High-voltage direct current has applications in some heavy industrial processes.

ELECTRICAL CIRCUITS

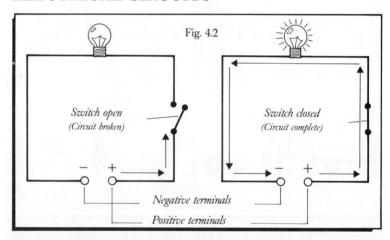

Fig. 4.2

Switch open
(Circuit broken)

Switch closed
(Circuit complete)

Negative terminals

Positive terminals

Figure 4.2 shows a simple electrical circuit, with a voltage supply (which may be mains voltage or a battery) having a positive terminal and a negative terminal. When the switch is closed, a current flows in the circuit from the positive terminal (+) to the negative terminal (–), lighting the lamp.

The lamp filament wire presents a resistance to the current flow and, as a result, the filament wire will heat up, giving out light. (The lamp is filled with a gas which makes the wire glow more brightly.)

In the case of a hairdryer, the 'element wire' heats up, a fan blows cool air over the hot wire, and hot air emerges from the nozzle to dry the hair (figure 4.3).

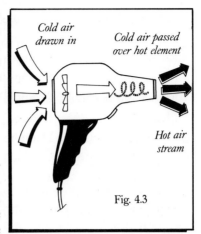

Fig. 4.3

FUSES

Faults in electrical equipment or wiring can have serious consequences, ranging from a mild shock to a serious fire. Because of these dangers, fuses should be incorporated in all electrical circuits. A fuse is a piece of wire which melts if a current greater than its 'rated current' flows through it. This breaks the circuit before the larger than expected current can do any damage — either to equipment connected to the circuit or to anyone using that equipment. After the fault has been traced and repaired, a new fuse is inserted and the power supply restored.

Apart from these mains fuses, electrical appliances also have a cartridge fuse fitted into the plug (figure 4.4). If a fault occurs in the appliance, the fuse will 'blow', isolating the appliance from the rest of the circuit and preventing any further damage.

Fuses in plugs are usually small glass tubes with metal contacts at both ends, linked inside by a piece of fuse wire. They are eas-

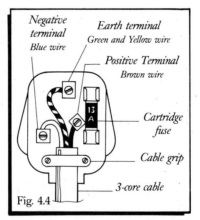

Fig. 4.4

ily replaced but it is absolutely vital that the correct fuse is fitted for each particular appliance. If in any doubt, this should be checked with an electrical dealer, or by referring to the manufacturer's operating instructions.

Some examples of fuse ratings and their applications

Electric shaver	—	1 amp
Table lamp	—	3 amps
Curling irons	—	3 amps
Hairdryer	—	5 amps ⎫
Iron	—	5 amps ⎭ up to 1000 watts

Vacuum cleaner	— 5 amps	up to 1000 watts
Refrigerator	— 5 amps	
Kettle	— 13 amps	
Hairdryer	— 13 amps	
Immersion heater	— 13 amps	1000 watts +
Electric fire	— 13 amps	
Washing machine	— 13 amps	

As well as having fuses, many appliances have an earth wire, which is connected to the earth terminal in the plug. If a fault occurs, the earth wire will carry the current away, preventing anyone handling the appliance from receiving a shock. (The current will, at the same time, blow the fuse and 'switch off' the appliance.)

Handled carefully, electricity is possibly the most convenient form of energy available. Handled without respect, it is a potential killer.

Some golden rules worth remembering

1 Never insert a plug into a socket, or remove it, with wet hands.

2 Never touch electrical appliances, or switches, with wet hands.

3 Ensure that the correct fuse is fitted.

4 Check the wiring on all appliances periodically to ensure that no bare wires are exposed.

5 Never replace a blown fuse until the fault has been traced and rectified.

TREATING A VICTIM OF ELECTRIC SHOCK

In the event of someone receiving an electric shock, the following procedures should be followed:

1 Turn off the power supply *immediately*. If you do not know where the mains switch is and the victim is still in contact with the faulty appliance, push the victim away from it using a wooden object such as a chair. Do *not* attempt to touch the victim because you will get a shock too.

2 Once the supply has been disconnected, and if the victim is unconscious, begin artificial respiration (mouth-to-mouth resuscitation) immediately.

3 Ask someone else to call an ambulance without delay.

SOME ELECTRICAL TERMS

Voltage (electrical pressure) is measured in *volts* (V).

Current (electrical flow) is measured in *amps* (A).

Resistance to electrical flow is measured in *ohms* (Ω).

Electrical *power* is measured in watts (W); it equals voltage multiplied by current.

Conductors are materials such as copper and iron, which will allow electricity to pass through them easily. For example, the wires in a three-core cable are made of copper.

Insulators are materials such as glass, porcelain, wood and plastic, which will not allow electricity to pass through them. For example, the wires in a three-core cable are covered with plastic sheathing to protect the user.

STATIC ELECTRICITY

Static electricity is created through friction when certain materials are rubbed together.

Continual brushing of the hair will often cause static electricity, making the hair become flyaway and difficult to manage. A distinct crackling can be heard when certain items of clothing are removed, particularly those made from synthetic fibres, and it is even possible to see sparks in a darkened room. Although a slight shock may be felt, static electricity is not harmful.

BATTERIES

Small electrical appliances such as torches and portable radios are powered by batteries. A battery, which is made up of individual cells, produces electricity by a chemical reaction between the substances which make up the positive and negative terminals of its cells. After a time, the chemicals lose their effectiveness and this type of *primary* battery has to be replaced by a new one.

Secondary batteries are rechargeable, however. A good example of a secondary battery is the type found in cars. The car battery provides the electricity to start the car, run the instruments and provide all the lights. When the engine is running, the battery is kept fully charged by an alternator. The chief drawback of this type of secondary battery is its weight.

5

Scalp massage

Whenever heat is introduced to the scalp, the blood supply will be increased. This is caused by the blood capillaries in the dermis dilating. As a result, more oxygen and nourishment are brought to the scalp and a more active hair growth is encouraged. Stimulation of the capillaries can be induced in three different ways, by using the hands, a vibro massager, or high-frequency apparatus.

HAND MASSAGE

There are two methods of massaging the scalp using the hands: *effleurage* and *petrissage*.

The effect of effleurage is to improve circulation in the veins and to increase the flow of blood towards the heart. Petrissage helps to remove inflammatory products contained within the tissue and increases the circulation.

Before starting any form of massage, the scalp must be checked for signs of soreness, cuts, bruises or irritation. It is unwise to proceed with treatment if any abnormalities are detected. It is also important that before starting, the operator's hands are washed thoroughly with an antiseptic soap and are completely dried.

To carry out the effleurage massage, stand behind the client and place the open palms of your hands on the forehead area of the scalp and using an even, but firm pressure, draw the hands back across the scalp, down the back of the head and into the neck, finishing at the shoulders. This action should be continually repeated for a period of 5 minutes.

Petrissage massage can be carried out on its own or used following a high-frequency treatment. After thoroughly washing and drying your hands, stand behind the client and place the tips of your fingers on the scalp with the thumb acting as a pivot. Using a firm, but comfortable pressure, the fingertips are moved in small circular movements, which prevent them from slipping over the skin, and by squeezing and stretching, the tissue under the fingers is moved. The massage is carried out for 5 minutes, avoiding the areas around the temples, following which time a pinkiness of the scalp, known as hyperaemia, will be noticed indicating that the massage has been effective.

VIBRO MASSAGE

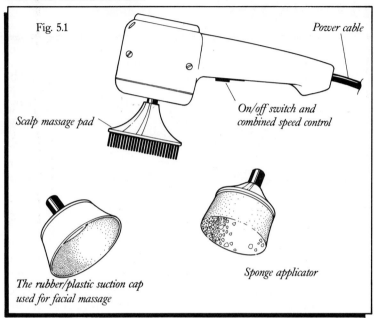

Fig. 5.1

Power cable

On/off switch and combined speed control

Scalp massage pad

The rubber/plastic suction cap used for facial massage

Sponge applicator

The vibro massager is a piece of hand-held electrical apparatus in which a small motor causes a shaft to vibrate (figure 5.1). A special rubber applicator is attached to this shaft and when moved over the scalp in circular movements, it creates friction, which in turn stimulates the blood supply.

HIGH-FREQUENCY MASSAGE

There has been no medical evidence to support the theory that high-frequency treatment encourages hair growth. It does, however, stimulate the blood supply to the scalp (the benefit of which has been previously explained). The apparatus used is of the box type (figure 5.2) and it has three interchangeable electrodes.

A low-amperage, high-frequency current is passed along a cable from the box to an electrode, which is inserted into a handle. Two controls will be found on the box: an on/off switch and a knob to control the strength of the frequency.

Before starting treatment, consideration must be given to the following points: the scalp must be clean, dry, and free from any form

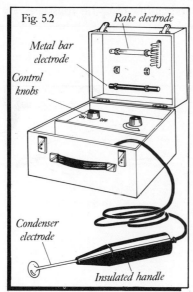

Fig. 5.2

Rake electrode

Metal bar electrode

Control knobs

Condenser electrode

Insulated handle

of spirit solution; both the client and operator should remove all metal items from their bodies. It is also important that the client is away from any metal fittings or wet surfaces.

The three electrodes are the rake or comb, the condenser, and the metal bar or saturator.

The *rake* or *comb* electrode, as the name suggests, is shaped like a rake and made of glass. It is secured in the handle and the machine is switched on with the frequency control set to a point just below medium strength. The handle is held by the operator in the right hand and two fingers of the left hand are placed on top of the electrode. The electrode is then placed on to the scalp where treatment is to start and the two fingers are removed. Using a gentle, steady movement, and taking care that the electrode remains in contact with the scalp at all times (the client will feel a slight shock if the electrode is not kept in contact with the scalp), treatment is continued for 5 minutes; then the two fingers of the left hand are once again placed on the electrode before removing it from the scalp. Throughout the treatment, the client should be able to feel the strength of the high frequency but without it being unbearable. Always ensure that she is feeling no discomfort and do not cause discomfort by lingering in the area of the temples.

The *condenser* electrode is a glass tube with a flat bulb at one end, and apart from the strength of the applied high frequency being set to a lower level than that used with the rake electrode, the procedure is the same. Whichever of these two electrodes is used, it is advisable to begin treatment at the crown of the head, working outwards towards the hairline and then back to the crown.

The *metal bar* electrode is a straight, metal rod, sometimes referred to as a saturator. Its method of use is quite different to that of the other two electrodes: after insertion into the handle it is given to the client to hold — the handle in one hand and the electrode in the other hand. To prevent the client from receiving a mild shock, a reasonably tight grip must be maintained throughout the treatment. The operator places the fingers of his left hand on to the scalp with a firm but comfortable pressure. After turning the machine on to a low frequency, the fingers of the right hand are also placed on the scalp and massage is carried out for 5 minutes with the fingers remaining in contact with the scalp at all times to prevent the client from receiving a mild shock. A tingling sensation will be felt by the operator when using this method. On completion, one hand only is removed from the scalp and the machine turned off. The electrode is taken from the client and a petrissage massage is given for a further 5 minutes.

6

Salon hygiene

SALON CLEANLINESS

Keeping a salon spotlessly clean and tidy is a never ending occupation. From the moment the doors are opened in the morning to their closure at night, there is always a floor to be swept, towels to be washed, mirrors to be polished, and many other jobs to be carried out. Not the most enviable task maybe, but certainly a very important one.

Clients have a right to expect their service to be carried out in clean, attractive and hygienic surroundings, and however clean and tidy the salon is before and following their appointment is of little consequence. It is *now* that matters.

Some points to remember

1 Hairs must be removed regularly from rollers, brushes and combs. They should all be washed frequently.

2 Trays used for holding rollers and permanent wave curlers must be cleaned regularly and any equipment used during colouring or bleaching processes must be thoroughly cleaned immediately after use.

3 Wipe around washbasins after each client, particularly after the removal of any chemical preparation.

4 Immediately clean chairs and work surfaces on which any preparation has been spilt.

5 Hair cuttings should be swept away as soon as possible, not only for the sake of appearance but also because they can easily cause a client or fellow staff member to slip, particularly on a tiled surface.

6 Do not leave wet towels lying around the salon. Not only is this untidy, but the smell can be offensive.

VISUAL DISPLAY

It is difficult to specify all of the factors which contribute to visual display, but basically this covers everything that the client sees during her visit to the salon. All of the factors already listed play a very important part, as do neat and tidy displays of such preparations as shampoos, setting lotions, etc. Clean, properly labelled bottles, carrying reputable brand names, inspire confidence.

Perhaps an easy way to appreciate what visual display is all about, is to take a look at the salon at the end of the day when all of the cleaning and tidying up has been carried out and everything is neatly in its place, ready for starting the next day's work. It is impractical to expect this situation to be fully maintained throughout the working day, but the aim should always be to match it as closely as possible.

STERILIZATION

The word 'sterile' means totally germ free. It is not possible or necessary for equipment and utensils used in a salon to be clinically sterile. Perfectly adequate cleanliness can be achieved by following these few simple guidelines:

For general sterilization, the regular use of a detergent and water will be sufficiently effective. Glass and porcelain objects can be cleaned by frequent washing in hot water, again with the addition of a detergent.

Metal instruments should have any surface dirt removed, followed by immersion in boiling water for a period of 20 minutes, or be placed in an automatic sterilizing cabinet. For towels, gowns and capes, regular hand or machine washing with the addition of soap powder will ensure that they are adequately sterilized.

On a daily basis, all work surfaces should be cleaned using sterilized cloths and a proprietory brand of cleaner, paying special attention to those hidden areas such as skirting boards and central heating pipes.

AUTOMATIC STERILIZERS

Two types of sterilizing cabinet are in common use: one relies on the vaporization of a formaldehyde solution to ensure effective sterilization of equipment, and the other uses ultraviolet (UV) rays.

The vaporization method — figure 6.1
The unit is usually a white plastic cabinet, approximately 600 mm (2 ft) high, 450 mm (18 in) wide, and 225 mm (9 in) deep. It has a glass or transparent plastic door and three shelves on which to place the equipment to be sterilized. These shelves are peppered

Fig. 6.1

The La Reine automatic sterilizer

with holes to allow the vapour to rise and circulate throughout the cabinet. The vaporizing unit is situated in the base of the cabinet. It consists of a heater which comes into contact with a container holding a sponge pad, on to which a formaldehyde solution is poured. When heated, the solution vaporizes, destroying germs on any equipment within the cabinet.

Method

1 Moisten the sponge with sterilizing fluid. Do not overmoisten. It is only necessary to use a few drops.

2 Remove surface dirt and hairs from the equipment to be sterilized, and place the equipment on the shelves.

3 For complete sterilization it is necessary to leave the instruments inside the cabinet for 30 minutes, during which time the door must remain closed.

The ultraviolet method

Ultraviolet (UV) rays provide a very effective method for destroying bacteria, and UV sterilizers have the added benefit of requiring no preparation before use, neither do they give off any odours. UV lamps are fitted inside the cabinet and require no more attention than to be switched on at the start of each working day and switched off at the end. This type of cabinet has a metal door so that the rays are contained within it.

Method

1 Place the items to be sterilized inside the cabinet and switch on, after having first removed any surface dirt and hairs.

2 For complete sterilization it is necessary to leave the instruments in the cabinet for 15 to 20 minutes.

3 A safety switch will automatically turn the unit off when the door is opened.

PERSONAL APPEARANCE AND HYGIENE

Hairdressing is a profession which offers its clients a personal service. It is also part of the fashion industry, and as with any other division of the fashion and beauty business, it is judged a great deal by the appearance of the people employed within it.

Some points to remember

1 Clothing, including overalls, should be clean, smart and tidy, and free from all stains.

2 Shoes must be clean and smart, and more importantly, comfortable enough to wear all day.

3 The otherwise perfect appearance of a female staff member can be spoilt by laddered tights and stockings.

4 Make-up should be applied carefully, and it goes without saying that hair must be neat, tidy, and attractively styled.

5 Because of the close physical contact the hairdresser has with a client, the personal hygiene of the hairdresser has to be beyond reproach. Even in the coldest weather, our bodies lose around 0.75 litres ($1\frac{1}{2}$ pints) of water every day in perspiration. Needless to say that in the atmosphere of a salon, this amount can be greatly increased. Regular and thorough washing, plus the use of deodorants, can save a lot of embarrassment.

6 Special attention should be paid to hands and fingernails. Hands should be clean and not stained by colouring products, and nails need to be clean, well-shaped and kept reasonably short.

7

Shampoo and shampooing

WHAT IS SHAMPOO?

Shampoo is a detergent, by which we mean a cleaning agent. The purpose of shampoo is to remove various types of dirt and grease from the hair and scalp, for example oils, body secretions and dust.

Hair that has been shampooed styles more easily and successfully than hair that has been dampened. The reason for this is that detergents encourage the cortex to absorb water, which causes the hair fibres to stretch far more easily during styling.

Two types of shampoo are manufactured: those which contain soap and those which are soapless. Shampoos containing soap are little used today because they have a number of disadvantages. They are alkaline, and therefore sting the eyes, dry the skin and do not form a lather easily in hard water.

Shampoos containing soap are produced by blending together an oil with an alkali. Both animal and vegetable oils can be used, such as whale oil or coconut oil.

Soapless shampoos are synthetic detergents, which as the name suggests, contain no soap. Scientists have developed chemicals which have the beneficial cleaning properties of soap but without the disadvantages, for example, they will form lather in hard water.

Soapless shampoos are produced by combining an oil, such as castor oil or mineral oil, with concentrated sulphuric acid. This produces a sulphonated oil. This is then mixed with an alkali which has a neutralizing effect. Good quality shampoos manufactured in this way will be slightly acidic (just as hair is), which is obviously beneficial. Another ingredient usually added to shampoo is a foaming agent — alkylolamide — and although it does not increase the cleansing action of the detergent, a rich foam is something which, from a psychological point of view, people have come to expect as being necessary.

The cleansing action of soapless shampoos is extremely effective. It is important to remember that unless a shampoo has been manufactured ready for use, it should always be diluted according to the instructions, to prevent it stripping away too much of the hair and skin's natural oils.

TYPES OF SHAMPOO

The basic types of shampoo designed for everyday use in the salon have a wide variety of additives which have limited benefits when used on the hair. It would not be reasonable to expect an anti-dandruff shampoo, for example, to cure such a condition; its main advantage would be that the condition would not be aggravated by its use.

More specialized, and consequently far more expensive shampoos, are available. With regular use they can be quite benefical to the hair, particularly when used as part of an intensive treatment programme.

Type	Usage
Base	— Normal hair and before permanent waving
Cream	— Normal to dry hair
Oil	— Dry hair
Lemon	— Normal to oily hair
Medicated	— Hair containing scurf

BENEFITS OF A GOOD SHAMPOO

1 It must be easy to spread through the hair, and lather easily.

2 It should rinse away easily and not leave a residue.

3 It should be of a neutral or slightly acidic pH.

4 It should enable the hair to be combed through easily.

5 It should not strip the hair totally of natural oils.

6 It should not make the hair become flyaway and difficult to manage.

SHAMPOOING TECHNIQUE

Shampooing should be looked upon as a service in its own right and not as an insignificant part of the whole treatment.

Carried out correctly, the shampoo stage is for many clients the most relaxing and enjoyable part of their visit to the salon. As such, the relevance of this service in relation to the overall success of a salon, should be obvious.

Method

1 Ensure that the client is protected by a gown, shampoo-cape and towel.

2 After sitting the client at the shampoo-bay, check that she is positioned comfortably.

3 Assess the hair and scalp condition and choose the correct shampoo accordingly.

4 Brush through the hair to remove any tangles.

5 Set the water to a comfortable heat and test on the back of your hand.

6 Soak all of the hair thoroughly, at the same time checking with the client that the water is at a comfortable temperature.

7 Apply the shampoo evenly over the head and, using the tips of the fingers, massage in a set pattern to ensure that every part of the head is covered.

The shampoo may not form a lather during this first shampoo stage — this will not affect the cleaning action.

8 Rinse the hair and apply a second amount of shampoo, repeat stage 7, and rinse very thoroughly.

When dealing with a greasy scalp condition, a slightly different shampooing technique is employed, the difference being as follows:

a) The water temperature is set to a cooler level as hot water tends to stimulate the sebaceous glands.

b) The massage is carried out more gently as again the sebaceous glands are stimulated by firm pressure.

Always explain to the client your reason for adopting this method of shampooing. Psychologically, a strong massage and hot water give the client the impression that this is the best way of tackling her greasy scalp condition. By explaining the shampooing technique you are going to use in advance, you will not only be making her aware of the advantages to be obtained by this method, but she will also be aware that you are knowledgeable in your craft.

When the hair is wet, harsh rubbing in any manner is harmful to it. After shampooing, excess moisture should be removed from the hair by blotting it between the folds of a towel, not with a vigorous rubbing action.

8

Conditioning the hair

Surprisingly, approximately 80% of the population have some form of hair or scalp disorder. The most common problems are the hair and scalp being too dry, too greasy, or having an excess of scurf.

It is not unusual, particularly in cases where clients are receiving regular chemical processes such as perming and tinting, to find two problems on the same head. For example, the scalp and root area may be greasy while the middle lengths and ends of the hair may be dry and porous. In recent years, technical developments in the conditioning field have been tremendous. It was not so long ago that the hairdresser's only real aids for improving the condition of hair were wax-based products, hot oil treatments, and acid rinses. Today, a whole variety of advanced products are available, from surface acting conditioners to sophisticated compounds which work at the hair's deep molecular level and on the scalp itself.

It is accepted that hair in good condition should be easy to manage, have 'bounce', movement, and shine. Hair in poor condition lacks these attributes, and consequently does not respond favourably to any chemical processes, such as permanent waving or colouring.

ASSESSMENT OF THE HAIR AND SCALP

A careful assessment of both hair and scalp is vitally important in order to establish whether there are any problems, and if there are, to determine the contributory factors and necessary action to correct the problem. This assessment can only be made on dry, unwashed hair.

The points to look out for are:

1 **Scalp condition**

Dry − greasy − scurf − dandruff.

True dandruff is a fungal infection of the scalp where there is an excessive production of sebum by the sebaceous glands. It is not very common. It can be recognised as thick, sticky, yellow bordered scales on the scalp, and is usually accompanied by spots around the hairline, particularly in the nape of the neck. Scurf (false dandruff) is easily recognised as loose, grey or whitish scales, which fall on to the shoulders when the hair is brushed.

Greasy and dry scalp conditions are easily recognised. The skin of a dry scalp tends to feel tight, whereas a greasy scalp is more pliable and can be moved by fingertip massage.

2 Hair condition

Dry — dry on the ends, greasy at the roots and greasy through the middle lengths — greasy throughout the entire length — porous and dull owing to chemical processes, sunlight, etc.

Apart from this visual check, it is also necessary to ask the client for information. For example, how long is it since her hair was last shampooed? It would be no use at all to go through an assessment procedure only to find on completion that the client had washed her own hair shortly before attending the salon.

Your opinion of the apparent condition of the hair and scalp should be made known to the client; she should also be allowed to explain any problems which she has experienced with her hair and/or scalp.

The environment in which we live, the conditions in which we work, our daily diet and many other factors, all have an important part to play in the condition of our hair and scalp. Other factors which can have an adverse effect include pregnancy, nervous disorders, drugs, and general ill-health.

SURFACE CONDITIONERS

A surface conditioner is designed to act as a lubricant on the hair shaft, allowing the comb to slide through the hair, and enabling tangles to be removed more easily. It makes the cuticle lie flat and leaves a fine film along the hair shaft, which reflects the light and gives the hair a shine.

Surface conditioners should be used on hair which is naturally dry, and following permanent waving, bleaching and tinting.

Some cheap surface conditioners may have a wax or mineral oil base and are best avoided as they tend to leave the hair sticky and lank.

Application

For a surface conditioner to work effectively, the correct application procedure must be followed.

After shampooing the hair, it should be towel dried. It is totally useless to apply a conditioner to hair which is dripping wet. Distribute the conditioner evenly throughout the head and comb right through the hair's length, using a large-toothed comb. The product is then left on the hair for the period specified by the manufacturer, followed by very thorough rinsing.

TREATING A GREASY HAIR AND SCALP CONDITION

Where the problem is one of excessive greasiness, there is no permanent cure.

Sepecialized products are now available to the hairdresser, and are most effective in many cases. Advice on these is best sought from the leading manufacturers of hair cosmetics.

The correct method of shampooing and choice of shampoo is important (see p. 31), as is a well-balanced diet. The client should be advised not to eat too many fried or fatty foods, or those containing chocolate.

RE-STRUCTURANTS

When hair is damaged at its molecular level, possibly by excessive chemical processes, a surface conditioner will only have the effect of masking the problem, and will be of no remedial value.

To fully understand the way in which a re-structurant works, and the benefits which it gives, we must first look at what happens when the internal structure of the hair is altered.

The hair is made up of a series of intertwined chains of keratin. In turn, keratin is made up of large molecules of different proteins which are linked together by cystine linkages. It is these linkages which are weakened, and even destroyed, when subjected to chemical and sometimes physical processes.

Generally, the hair is capable of withstanding most processes but when it is abused, or subjected to a number of harmful factors in a short period of time, it will loose its elasticity, become porous and dull, and unreceptive to any treatment carried out. The more cystine linkages weakened or destroyed, the bigger the problem becomes. In severe cases, where a very high percentage of linkages have been destroyed, disintegration can occur. At this point, no help is available.

The purpose of a re-structurant is to penetrate the areas in the hair where the cystine linkages are weakened, in order to strengthen them. Having done this (a number of applications may well be needed) the hair will regain its elasticity and manageability, to the satisfaction of both client and stylist.

The effect which a re-structurant product has on damaged hair is shown in figure 8.1.

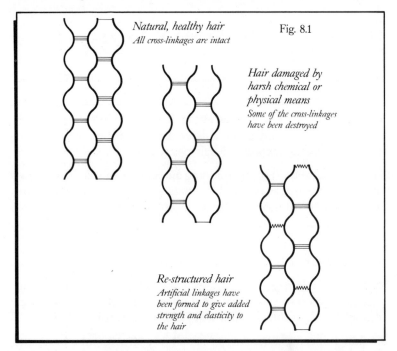

Natural, healthy hair
All cross-linkages are intact

Fig. 8.1

Hair damaged by
harsh chemical or
physical means
*Some of the cross-linkages
have been destroyed*

Re-structured hair
*Artificial linkages have
been formed to give added
strength and elasticity to
the hair*

It is unfair to expect miracles from any conditioning product, particularly in cases of severely damaged hair, but by making the correct assessment, and using the appropriate treatment for the problem concerned, a high standard of hair condition can be achieved on most heads of hair.

9

Permanent waving

Making straight hair curly is far from being a new idea. Even in the time of the Pharoahs, women wound their hair around wooden rods which were then covered in wet clay from a river bed and baked in the hot sun to form waves. This method may have had its disadvantages, but it is doubtful whether they suffered as much discomfort as those women undergoing permanent waving processes during the earlier part of this century. One popular method, which took many hours to complete, involved the use of elaborate electrical apparatus. It was most uncomfortable, and at times physically harmful. An alkaline solution would be applied to the hair and then, using the electrical equipment, heated to 100 °C after being wound on curlers. To give protection from such high temperatures, felt pads were placed between the scalp and wound hair. After sufficient processing time, which could be considerable, the alkali was removed by rinsing through the hair with a diluted acid solution.

COLD WAVING

It was in the United States, in the 1930s, that cold waving was introduced and this is now the most universally used method of permanent waving. Although acid-based solutions are available, the alkaline-based products still prove to be the most popular for general use. Thioglycollate is used as the base constituent, this being a combination of ammonium hydroxide and thioglycollic acid, and although no heat is applied to a cold wave, the chemical reaction requires the warmth of a comfortable room temperature to be most effective.

HOW A COLD PERMANENT WAVE WORKS

In very simple terms, cold waving is a controlled breaking down and re-building of the hair's structure using chemicals (figures 9.1 to 9.4). The keratin, of which hair is composed, is formed in long chains which are held together by a system of cross-linkages known as disulphide bonds. These are made up of a sulphur-containing substance called cystine and when the waving lotion is applied to the hair, a high percentage of these linkages are broken down, at which point they become cystéine. This action is stopped when water is applied.

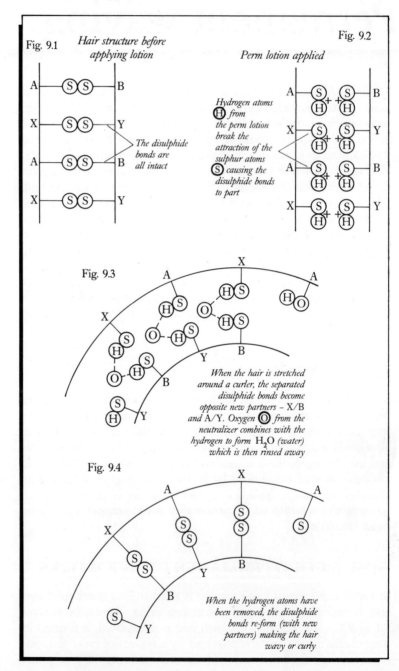

Fig. 9.1 *Hair structure before applying lotion*

The disulphide bonds are all intact

Fig. 9.2 *Perm lotion applied*

Hydrogen atoms Ⓗ from the perm lotion break the attraction of the sulphur atoms Ⓢ causing the disulphide bonds to part

Fig. 9.3

When the hair is stretched around a curler, the separated disulphide bonds become opposite new partners – X/B and A/Y. Oxygen Ⓞ from the neutralizer combines with the hydrogen to form H_2O (water) which is then rinsed away

Fig. 9.4

When the hydrogen atoms have been removed, the disulphide bonds re-form (with new partners) making the hair wavy or curly

An oxidizing agent is then used to re-form the linkages with different partners in new positions, to give a wavy or curly result. If the action of the lotion is allowed to carry on beyond the optimum point for a successful result, a depilatory action will occur, dissolving the hair completely. Modern cold waves are extremely versatile and winding methods can be adapted to produce results ranging from soft waves to tight curls. Processing times vary and are dependent on a number of factors such as temperature, type of lotion, hair texture and whether the hair is tinted or bleached. Unless the product being used has a fixed processing time, a test curl must be taken every 3 to 5 minutes.

There are permanent wave lotions manufactured for every hair type, therefore correct assessment is essential before selecting which one to use.

Some points to look for

1 Texture.

2 Condition and porosity. If there is any doubt about the hair being in a suitable condition, do not attempt permanent waving.

3 Check whether the hair is tinted or bleached, or double processed (that is, tint over the top of bleach) in which case a lotion for bleached hair would need to be selected.

Another point to bear in mind is that the hair is a modified extension of the skin. Any lotion which will break down the hair structure will also damage the skin. Unless otherwise stated by the manufacturer, permanent wave lotions should be applied to each mesh of hair immediately before winding and to within 12 mm ($\frac{1}{2}$ in) of the scalp. Great care must also be exercised when applying further lotion on completion of winding, to ensure that it does not flood on to the scalp.

Some points to remember

1 The size of curler determines the tightness of curl achieved.

2 A permanent wave gives best results on hair up to 225 mm (9 in) in length.

3 Never apply permanent wave lotion to a head where the scalp shows signs of soreness or abrasions.

4 Always wind curlers at right angles to the scalp (figure 9.5).

5 The size of section taken should be the same overall size as the curler being used.

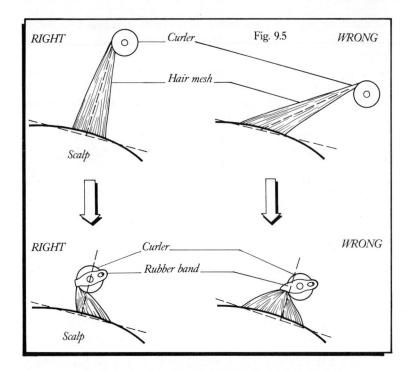

Fig. 9.5

RIGHT — Curler — Hair mesh — WRONG

Scalp

RIGHT — Curler — Rubber band — WRONG

Scalp

COLD PERMANENT WAVING METHOD

1 Assess the condition of the hair and select the correct lotion. Ensure that the scalp is free from soreness and abrasions.

2 Shampoo the hair with a mild, base shampoo, and cut into the desired style. The hair should be left damp for winding unless otherwise stated by the manufacturer.

3 Ensure that the client is adequately protected by gown and towels from the risk of dripping lotion.

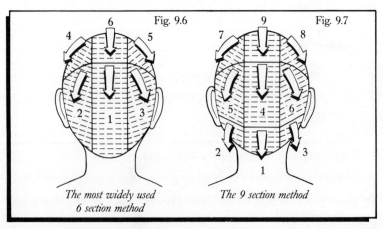

Fig. 9.6

Fig. 9.7

The most widely used 6 section method

The 9 section method

4 The winding technique will follow either a conventional pattern (figures 9.6 and 9.7), or be directional. A directional wind entails positioning the curlers in the direction of the intended style, in much the same way as one would set a head of hair on rollers.

5 For a conventional wind, start at the crown of the head (figure 9.6 section 1) and take a mesh of hair ensuring that the section is no wider than the curler being used. Hold the hair at right angles to the scalp and comb through from roots to points. Apply lotion to within 12 mm ($\frac{1}{2}$ in) of the roots and place a curler underneath the ends of the hair so that the hair lies flat on the curler. Bunching of the hair in the centre of the curler must be avoided.

6 Place an end-paper on top of the hair, so that it protrudes a little way beyond the ends. This will ensure that 'fish hooks' (badly damaged ends) are avoided.

7 With an even tension, wind the curler down to the scalp, the end-paper being wound around the curler first. Fasten the rubber across the curler making sure that it is not twisted or pressing down on to the roots of the hair. This could cause severe marking of the hair, or at worst, breakage.

8 Continue winding as quickly as possible, at the same time taking care to wind correctly, until all the hair has been wound.

9 Place a cotton wool strip around the entire hairline and re-dampen each curler with lotion. Do not saturate.

10 Replace the cotton wool with a fresh, dry strip.

11 Cover the head with a plastic cap. This helps to retain the heat from the scalp and prevents evaporation of the lotion.

12 After no more than 5 minutes, take a test curl. This is done by unwinding a curler a little way from the scalp and, with the hair held against the curler by the flat part of the thumbs, it is pushed forward towards the scalp. A satisfactory result has been achieved when the hair falls into an 'S-shaped' wave. On average, cold waves require a processing time of 10 to 15 minutes on normal hair and 5 to 10 minutes on tinted or mildly bleached hair.

13 Move the client to a shampoo basin and rinse the wound curlers for a full 5 minutes, using warm water.

14 Place a towel over the head and gently blot away all excess water. Use two towels if necessary as the hair must only be damp when the neutralizer is applied.

15 Apply the appropriate neutralizer to the entire head, being careful to ensure that every curler is well saturated, and leave for

5 minutes. Some neutralizers need to be poured into a bowl and foamed before application, whereas others are applied directly from an applicator bottle.

16 Remove the curlers gently and apply more neutralizer, working it well into the ends. Leave for a further 5 minutes and rinse thoroughly with warm water. The hair is now ready for setting or blow-drying.

Apart from cold waves, there are a number of other methods for permanently waving hair, although in virtually every case, chemicals are used. Some require the addition of heat, either from an external source such as a hood-dryer, or by using heated clamps, while exothermic permanent waves generate their own heat when two chemical solutions are mixed together.

MACHINE PERMING

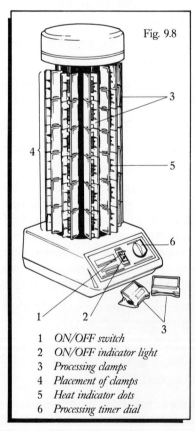

Fig. 9.8

1   ON/OFF switch
2   ON/OFF indicator light
3   Processing clamps
4   Placement of clamps
5   Heat indicator dots
6   Processing timer dial

This is an old idea brought up to date and is proving to be very popular. The method used is relatively simple and removes much of the risk of human error. Illustrated in figure 9.8 is the very successful UniPerm machine system marketed by Clynol Hair.

How the system works
The primary function of the machine is to heat special processing clamps to a pre-determined temperature. These are then applied to the wound hair on which a specially formulated lotion has been used. The clamps release their heat over a period of 6 minutes, after which time processing is complete. The winding technique is different to that used in cold waving and should only be used in conjunction with machine systems.

Machine perming method

1 Ensure that all of the precautions usually taken before permanent waving are observed.

2 Select the correct size of clamp in relation to the size of curlers being used and fit them on to the bars of the machine. The clamps will require approximately 20 minutes to heat up and it is, therefore, important that the machine is switched on before winding begins.

3 Shampoo the hair with a mild, base shampoo, and towel dry.

4 Select the appropriate lotion for the hair.

5 Section the hair in the conventional pattern (see figure 9.6 p. 40) or, if directionally winding, in the line of the style.

6 Begin taking partings of 25 mm (1 in) at the nape section and thoroughly dampen each hair mesh to within 25 mm (1 in) of the scalp. Although the size of the sections is much larger than could be successfully used with regular permanent waving, the machine system is able to give a perfect curl pattern on sections of this size (figure 9.9).

7 Begin winding the hair, continually easing the mesh towards the centre of the curler so that the clamp will subsequently cover all of the hair (figure 9.10). Be sure that the hair is wound smoothly and securely, and placed directly in the centre of the section.

8 Continue winding in the same manner throughout the head.

9 Check that the heat-indicator dots on the clamps are dark brown in colour. This shows that the correct temperature has been reached.

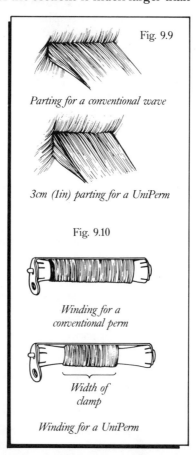

Fig. 9.9

Parting for a conventional wave

3cm (1in) parting for a UniPerm

Fig. 9.10

Winding for a conventional perm

Width of clamp

Winding for a UniPerm

10 Re-dampen each curler with lotion and adjust the rubbers so that they lie along the top of the curlers and away from the direct heat of the clamps.

11 After removing the clamps from the machine, immediately apply them on to the curlers (as shown in figure 9.11), and ensure that all of the hair is covered.

12 Process for the stated time. Do not take test curls, but remove the clamps and apply the neutralizer.

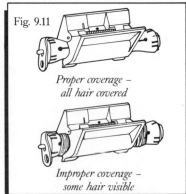

Fig. 9.11

*Proper coverage –
all hair covered*

*Improper coverage –
some hair visible*

STYLE SUPPORTS

The purpose of style supports is to give soft waves without definite curls, the emphasis being more on root lift. Their popularity has arisen with the advent of blow-dry styling techniques, for which they are exceptionally well suited. The hair is wound on rollers which are generally larger than permanent wave curlers, following the direction of the intended style. Style supports are effective for 6 to 8 weeks, about half the length of time normally expected from a permanent wave. Not only do they enable the stylist to achieve better results than can normally be obtained when styling very straight hair, but they also ensure easier manageability for the client in between salon visits.

RECOGNIZING FAULTS IN PERMANENT WAVING

Being able to recognize what has gone wrong is almost as important as knowing how to do something correctly. In permanent waving there are many faults which can occur, virtually all resulting from human error.

1 **An end result which is too curly**

(a) The curlers used were too small.

(b) Misinterpretation of required result.

2 **Hair straight when dry but curly when wet**

(a) Incorrect choice of permanent wave lotion. Too strong for hair type.

(b) Hair wound under too much tension.

(c) Over-processing (for example, the lotion was left on the hair for too long).

3 No result on completion

(a) Processing time was insufficient.

(b) Poor neutralizing procedure.

(c) Curlers used were too large.

(d) Insufficient lotion applied.

(e) Hair not adequately towel-dried before winding.

(f) Lotion too mild for hair type.

4 Uneven wave formation

(a) Not enough care taken in ensuring that the lotion was applied evenly to each section.

(b) Too much water left in the hair before winding.

5 Rapid relaxation

(a) Poor neutralizing. Insufficient neutralizer has been applied, or excess water has not been adequately removed following rinsing.

6 Hair breakage

(a) The lotion was too strong.

(b) Over-processing.

(c) Hair wound under too much tension.

(d) Rubbers on perm. curlers twisted or incorrectly positioned.

7 Soreness of the scalp during or following perming

(a) Lotion allowed to flow on to the scalp during winding. Often accompanied by too much tension on the hair.

(b) Cuts and abrasions had not been noticed or had been ignored.

TESTING FOR COMPATIBILITY

As long as the normal precautions have been taken, permanent wave lotions can be used on most heads with perfectly successful results. A major problem can arise however if products containing metallic salts have previously been applied to the hair. These can be in the form of compound hennas, hair colour restoratives, or coloured hair sprays. The fact that metallic salts are contained in a product is not always specified by the manufacturer. Hair breakage can arise if permanent wave lotions are applied to a head treated with such preparations, so a compatibility test should *always* be carried out where there is any doubt.

Method

1 Take a number of small hair cuttings from all over the head and tie them together at one end using a piece of cotton.

2 Immerse them in a small amount of permanent wave solution for 15 minutes.

3 Rinse thoroughly. Blot out any excess moisture from the cuttings and place them in neutralizer; make sure it contains a percentage of hydrogen peroxide.

4 Leave for 10 minutes and then examine them for signs of damage or weakness. If all is well, proceed with permanent waving.

In the event of metallic salts being present, wisps of smoke may be seen rising from the clippings when they are placed in the neutralizer.

10

Straightening hair

There are two occasions when it may be necessary or desirable to straighten hair. Firstly, if the hair is naturally too curly, therefore preventing the stylist from successfully creating the style required by the client; secondly, when a head of hair is permanently waved too tightly and some relaxation is required. Many hairdressers are reluctant to carry out a straightening process, more often than not because of tales of disastrous results, which may be based on myth rather than personal experience. There is absolutely no reason why hair should not be straightened successfully if the hairdresser is fully experienced and if care is taken during the treatment.

STRAIGHTENING NATURALLY CURLY OR FRIZZY HAIR

The best method for straightening naturally curly or frizzy hair is to use a product designed specifically for this purpose. Permanent wave lotions are not very sucessful; the main problem is that their thin consistency prevents the hair from being held straight during the processing period. Continual combing to combat this can be extremely damaging. Hair straighteners on the other hand, have a thick consistency and hold the hair firmly in place. They are available either as creams or a liquid and powder mixture.

In general, the method of application is as follows:

The straightener is applied with a tinting brush to unwashed hair. Starting at the nape, take 12 mm ($\frac{1}{2}$ in) sections and apply the cream along the entire length of the mesh of hair with the exception of the first 12 mm ($\frac{1}{2}$ in) at the roots. Work on both sides of the head at the same time to ensure even processing and on completion ensure that the hair is lying flat and is not in contact with the client's skin. Make the application as quickly as possible, but at the same time be thorough. Pay particular attention to the processing time given by the manufacturer.

The next stage is also of critical importance. Rinse the hair with warm water for a full 5 minutes, lifting it gently from the scalp to ensure that not a trace of the straightener is left in the hair. It is important that the hair is blotted in the folds of a towel, not rubbed, until as much water as possible has been removed. Apply the neutralizer, lifting the hair as you do so to ensure total penetration; leave for the specified time and rinse very thoroughly.

STRAIGHTENING PERMANENTLY WAVED HAIR

Hair which has been permanently waved is usually very receptive to being straightened or to having the curl relaxed for a softer result.

Method

Mix very thoroughly 57 g (2 oz) of permanent wave lotion with one phial of conditioning cream (see p. 39 — Some points to look for). Apply the lotion to pre-dampened hair in 25 mm (1 in) sections, starting at the nape. Use a sponge and avoid lotion being applied to the scalp. For the sake of convenience and client comfort, this procedure is best carried out with the client seated at a backwash basin. On completion, comb through once or twice only — continuous combing will put a strain on the hair and may lead to breakage. Straightening permanently waved hair is usually a fairly quick operation, so a check should be made every minute until the required degree of relaxation or straightness has been achieved. Rinse the head thoroughly for 5 minutes, gently lifting the hair to ensure that the rinsing water penetrates through to the scalp. Blot out excess moisture with a towel and apply the appropriate neutralizer, again ensuring total penetration. Leave for 10 minutes, rinse well and apply a conditioner.

Hair straightening should not be carried out if there is any doubt about the hair's condition being suitable. It is never possible to guarantee total straightness, a point which should be made clear to the client before starting treatment.

11

Hair colouring and bleaching

Imagine how drab and dull the world would be without colour — a world where everything appeared in black and white, and varying shades of grey. The startling contrast can be seen by comparing a photograph or film produced in black and white with the same subject produced in colour. We accept colour as part of our everyday lives, usually without giving a second thought to the effect it has on us.

Perhaps the first people to take full advantage of the subtle yet marked effect colour has on our everyday lives, were those involved in the world of advertising. With the coming of supermarkets and self-selection of goods, entirely new methods for attracting people to buy particular products had to be found. The one which made the biggest impact was correct packaging. Early experiments soon proved that a product displayed in a colourful, interesting package, would sell in far greater volume than the identical product contained in a plain, uninteresting package.

Make-up has been used for centuries, by both women and men, to enhance features and give a more attractive appearance, and hair colouring itself also dates back many centuries.

Hair colouring can only be fully successful if it is completely understood. Although modern colouring preparations are very advanced, and the instructions for their use are relatively simple, a thorough technical knowledge is necessary to ensure successful results.

UNDERSTANDING COLOUR

White light (full sunlight) is made up of all colours — figure 11.1 shows how these colours of the rainbow can be seen by shining white light through a glass prism. The colours are red, orange, yellow, green, blue, indigo and violet.

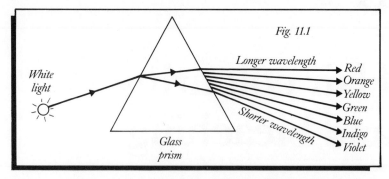

Fig. 11.1

White light

Glass prism

Longer wavelength

Shorter wavelength

Red
Orange
Yellow
Green
Blue
Indigo
Violet

Objects only look coloured because of their ability to absorb some colours from the light shining on them and to reflect the other colours — giving the colour we see the object to be. If the object reflects all colours, it appears to be white. If the object absorbs all colours (or if no light shines on it) it appears to be black.

Disregarding indigo (as being difficult to distinguish from either blue or violet), we are left with six colours: red, yellow and blue, and orange, green and violet. Red, yellow and blue are called the *primary* paint (or pigment) colours; orange, green and violet are called the *secondary* paint colours.

If we mix two primary paint colours, we produce a secondary colour:

$$red + yellow = orange$$
$$yellow + blue = green$$
$$red + blue = violet$$

Figure 11.2 shows two superimposed triangles — one with the primary paint colours and one with the secondary paint colours. The colours that are opposite each other in this star shape will neutralize each other:

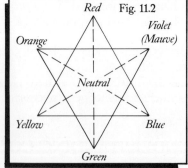

Fig. 11.2

Red
Violet (Mauve)
Orange
Neutral
Yellow
Blue
Green

red neutralizes green
violet neutralizes yellow
blue neutralizes orange

Example: A client's hair contains too much yellow, following a mild bleach. To remove this unwanted yellow, it would be necessary to use a tint which contained a violet element. If a tint were to be applied which contained one of the other colour elements, a problem would arise. For example, a blue-based colour, applied to the same hair, would give a 'green' result, because blue + yellow = green.

All manufacturers of permanent tints have a coding system, usually in the form of colour charts, which explains both the colour depth and tone element of their products. Further information is readily available from the manufacturers.

DEPTH AND TONE

Colour can be split into two parts: depth — how light or dark — and tone — the actual colour (gold, warm, ash, etc.) (figure 11.3). Natural hair colours all have a depth and usually a visible tone. Where no obvious tone is apparent, it is referred to as a basic shade. The greater the intensity of a colour, the stronger the colour to neutralize it will have to be. It would be of no use for example, expecting a vibrant yellow to be neutralized by applying a very pale violet. A far greater concentration of violet would be necessary.

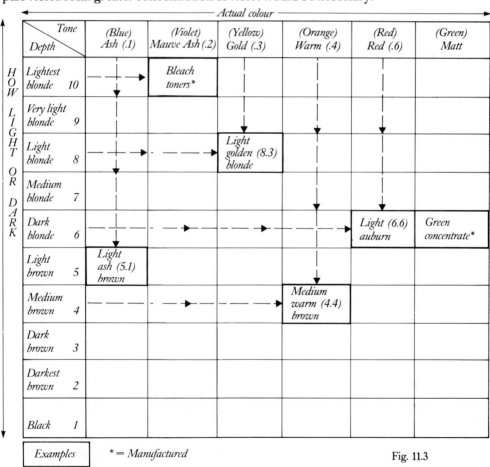

		Actual colour					
	Tone / Depth	(Blue) Ash (.1)	(Violet) Mauve Ash (.2)	(Yellow) Gold (.3)	(Orange) Warm (.4)	(Red) Red (.6)	(Green) Matt
HOW LIGHT OR DARK	Lightest blonde 10		Bleach toners*				
	Very light blonde 9						
	Light blonde 8			Light golden (8.3) blonde			
	Medium blonde 7						
	Dark blonde 6					Light (6.6) auburn	Green concentrate*
	Light brown 5	Light ash (5.1) brown					
	Medium brown 4				Medium warm (4.4) brown		
	Dark brown 3						
	Darkest brown 2						
	Black 1						

Examples	* = Manufactured	Fig. 11.3

Whenever hair is lightened, either by a tint or bleach, the red and yellow pigments which it contains are accentuated. These tones may be desirable where a warm or golden result is required, and therefore it may be unnecessary to use a neutralizing shade, in fact a basic shade could achieve the target colour (the end result required). For example where a target shade of light warm brown is required on a medium brown natural base, this result could well be achieved by using a basic shade which is lighter than the natural base, the warmth coming from the red pigment contained within the hair. If a light warm brown shade of tint were to be applied, it is possible that the end result would be too red, because we would be adding colour to colour.

The effect which red and yellow pigments have when hair is lightened, is explained more fully in the chapter on bleaching.

THE EFFECT OF LIGHT ON HAIR COLOUR

The colour of a head of hair will appear different under varying light sources. The truest result will be seen in normal daylight. Under tungsten lamps or spotlights, the colour will appear warmer, because these lights contain more red. Fluorescent lights contain more blue, so making the colours appear more ashen.

It is important when working with colour in the salon, to work in daylight wherever possible. If this is not possible, the ideal situation is a mixture of tungsten and fluorescent light.

The effect of light on hair colour should be explained to a client having a tint for the first time, or a client who is changing to a different shade, and wherever possible, allow her to select a shade from the colour chart in daylight.

THE PRE-DISPOSITION TEST (SKIN TEST)

There is a chance that a client may be allergic to a tinting product, so a pre-disposition test should be carried out on any client who has not received a tint application for a period of three months or more. There are two reasons for this. Firstly, to establish whether the skin has an abnormal sensitivity to hydrogen peroxide; secondly, and far more importantly, whether the client is allergic to the para-chemicals contained in the tint.

An allergy to the latter can be most serious, bringing about a severe case of dermatitis, which in the extreme case can cause vomiting, fainting, running sores, swelling around the eyes, and even temporary blindness. Severe cases like this are not common, but the risk, however slight, is definitely not worth taking.

The test is simple to carry out, but will necessitate the client attending the salon at least 48 hours before the intended application is made.

Method

1 Clean an area behind one ear, preferably with surgical spirit, and apply a small amount of dark tint plus peroxide to this area. The para-phenylenediamine content of darker tints is more likely to show any abnormal sensitivity.

2 Allow the tint to dry, and explain to the client that should she feel any irritation or discomfort from the tested area within the next 24 to 48 hours, she should wash the area thoroughly with soap and water. She should also notify the salon.

3 Under no circumstances should a tint application be made to a head where the result has proved positive.

Word of warning

Para-chemicals are not confined to permanent tints only, but may be found in semi-permanent colours. Consider this point when an application of permanent tint has been ruled out — the alternative choice of colouring product may also hold risks.

HAIR COLOURANTS

The practice of hair colouring is virtually as old as the human race, the first dyes being extracted from vegetable sources. The advancements made in the quality and the naturalness of hair colourants have taken place mainly since the 1950s, although it is to a small number of European chemists, in the late nineteeth century, that the credit should go for the basic formulations of modern permanent tints.

TEMPORARY COLOURS

As their name suggests, these colouring products will only last from one shampoo to the next.

They are applied to the hair after shampooing, either as a rinse or as a coloured setting lotion, adding delicate colour tones to the hair.

A temporary colour surrounds the hair cuticle, and has no power to lighten the natural hair colour, or bring about any drastic change.

SEMI-PERMANENT COLOURS

Semi-permanent colours are important for introducing colour to the client who is a little nervous of committing herself to a permanent tint. Considerably stronger than temporary colours, semi-permanents will darken hair, add strong colour tones, and blend in a reasonable percentage of white hairs, but they have no power to lighten the hair. Application is made on damp hair which has either been pre-shampooed or dampened with water, depending on the product used, and allowed to develop for between 10 and 20 minutes. Semi-permanent colours can be in a liquid, foam, or cream form, and the colour particles become lodged behind the cuticle, gradually being removed each time the hair is washed.

PERMANENT TINTS

Modern permanent tints are extremely versatile, being able to lighten or darken the natural hair colour, cover 100% white hair, or just add tone.

Many shades are available, which the skilled colourist can use to add new dimensions to both a client's overall appearance, and to a hairstyle.

USING PERMANENT TINTS

Head assessment

Before choosing a hair colour for a client, many factors need to be considered, each one having an effect on the success of the final result.

Factors for consideration are:

1 **The hair's natural colour depth**

   Permanent tints can usually lighten the hair up to three natural shades, for example, from light brown to medium blonde. In certain cases, this can be increased to four shades when working on dark blonde or lighter hair. Hair which is medium or dark brown however, may not lighten more than two shades. By knowing the natural colour depth (how light or dark) and the target shade (colour required), we can establish whether the amount of lift needed can be achieved without pre-lightening.

2 **The natural colour tone**

   By knowing the colour tone of the hair, we can judge what reaction there will be to the lightening and predict the effect the tint colour will have.

3 Percentage of white hair

Where a high percentage of white hair is present, slight variations in technique may be needed. For example, the tint application should start at the crown, so that the top section of the head is completed first (see figure 11.4). Also, when using fashion shades, it will be necessary to mix them with an equal amount of base shade of an equivalant depth, to prevent over-exaggeration of the final colour; for example, very light auburn (6.6) would be mixed with the equivalent base shade − dark blonde (6).

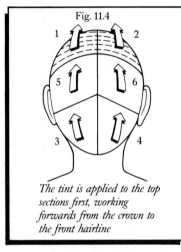

Fig. 11.4

The tint is applied to the top sections first, working forwards from the crown to the front hairline

4 Previous treatments

Hair which is already tinted cannot in general be further lightened by using another tint.

Hair which is bleached is usually more porous, and may tend to 'grab' the colour, the end result appearing darker than required.

Porous hair also allows colours to fade more rapidly, therefore it may be necessary to pre-pigment the hair (put in either a red or gold shade before the target shade is applied). Porosity may also need to be taken into account when making a colour application to permanently waved hair.

5 Condition

Hair which is in poor condition, particularly when brought about by other chemical treatments, should not be tinted. The client should be advised to have a course of corrective treatments until such time as the hair is strong enough to undergo an oxidation process.

6 Client requirements/limitations

Very often a client asks for a particular shade of colour because she has seen a picture, or another person, with that shade. As will be seen from the points given in section 7, this choice may not be at all suitable. There may be circumstances where it is not possible to give the client the colour required, because of certain limitations. For example, the condition of the hair, the need to pre-lighten with bleach being against the client's wishes, or the cost of the service necessary to achieve the desired result.

7 Skin colour and colour of eyes

These factors are allied to the points in section 6, and are mainly a matter of common sense. Lighter colours will be more suitable for the older client, and in many cases for people with light coloured eyes.

A colour tone which is complimentary to the skin tone should be chosen; exact opposites tend to look unnatural.

8 Scalp condition

Examine the scalp for any irritation or abrasions. Do not proceed with any chemical process where these problems are noticed.

HOW A PERMANENT TINT WORKS

The colouring action which takes place during tinting is the result of a chemical action between para-phenylenediamine or para-toluenediamine and hydrogen peroxide.

When a tint is applied to the hair, the ammonia in the tint softens and opens the cuticle. The natural pigment is oxidized, colour molecules penetrate into the cortex and combine to form larger molecules, which remain deposited in the cortex. Following a pre-determined processing time, the surplus amount is washed away.

By altering the balance of certain chemicals with the tint, different colour depths and tones can be achieved.

INGREDIENTS OF A TINT

Permanent tints are manufactured from synthetic organic dyes, derived from coal tar. They are as follows:

Para-phenylenediamine

Meta-phenylenediamine

Para-toluenediamine

Meta-toluenediamine

Para-aminophenol

Para-aminodiphenylamine

Diaminophenol

Diaminoanisol

A tint will also contain colour modifiers, for example, resorcinol. This stabilises the chemical process and regulates the speed of oxidation. Antioxidants (such as sodium sulphite) prevent the tint from darkening on exposure to air, and ammonia (ammonium hydroxide) adjusts the pH value, opens the cuticle, and acts as a catalyst to disengage the oxygen from the peroxide more rapidly.

A conditioning base, usually cream or oil, contains a detergent that acts as a wetting agent, enabling the tint to take more effectively.

TINT APPLICATION PROCEDURES

First time application
Hair that contains no form of artificial colourant is referred to as *virgin* hair.

Because of heat from the scalp, and the fact that complete keratinization of the hair may not have taken place at the root section, the first 25 mm (1 in) of the hair near the scalp will react differently to the tint, compared to the middle lengths and ends. When lightening virgin hair therefore, or when using a fashion shade, application is made first to the middle lengths and ends, leaving this 25 mm (1 in) of root section untouched.

When the tint has developed for 15 minutes, a fresh amount should be mixed, and applied to the root area. The full development time specified by the manufacturer should then be allowed following this second application.

When darkening virgin hair, the tint can be applied on the entire length of the hair, and a heavy application should be made.

Regrowth application (short)

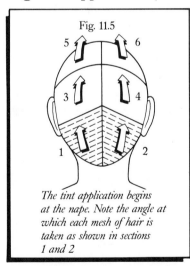

Fig. 11.5

The tint application begins at the nape. Note the angle at which each mesh of hair is taken as shown in sections 1 and 2

Apply the tint to the roots only, starting at the nape (figure 11.5). If the colour on the middle lengths and ends needs refreshing, mix a fresh amount of tint (using 10 volume peroxide) and brush on during the last few minutes of development.

Do not try to spread the tint from the roots on to the middle lengths and ends.

Regrowth application (long)
When darkening hair, a heavy application of tint can be made to the entire regrowth.

When lightening hair, treat the middle lengths first, as though working on virgin hair. For example, on a 75 mm (3 in) regrowth, the first 25 mm (1 in) near the roots would be left and the tint applied to the next 50 mm (2 in) only. After 15 minutes, fresh tint would be applied to the first 25 mm (1 in) and the full development period allowed.

REMOVAL PROCEDURE

The correct removal of a tint is of great importance.

When the full development time has been reached, move the client to a washbasin and sprinkle a small amount of water over the head. Give a gentle massage, paying special attention to the hairline area, particularly in the nape of the neck, and carry on until all of the tint has lifted from the scalp.

Rinse very thoroughly until the water runs clear and then wash the hair with an antioxidant shampoo (if available) or a mild base shampoo.

Some points to remember

1 Do not mix tints in metallic bowls – use those made from either glass or plastic.

2 Always mix the tint in the way described by the manufacturer.

3 As a general rule, tint will only lighten virgin hair.

4 Apply the tint neatly and carefully making sure you apply a sufficient amount to keep the hair moist. This will help to prevent staining.

5 Start at the nape section when applying tint unless there is a high percentage of white hair at the front.

COLOUR STRIPPING

In certain circumstances it is necessary to remove artificial colourants from the hair. The process is referred to as *colour stripping*, for which special products are manufactured.

Some instances when a colour stripper would be used are:

1 On a head of hair which has been tinted to a shade darker than the one required.

2 When a client wishes to change the colour of her already tinted hair to a lighter shade.

3 Where streaks or meshes are required in tinted hair, before bleaching.

Colour strippers will remove semi-permanent and permanent colours from the hair. They do this through stages of red and yellow in much the same way as a bleach, which should be explained to a client before the process is started, because it can appear to be drastic.

Because the method can vary between different products, it must be emphasized that before using any colour stripper, the instructions must be read and followed implicitly.

Method

1 Establish the type of colour in the hair, permanent or semi-permanent, and mix the colour stripper accordingly.

2 Apply to the darkest or most resistant areas first, and then to the lighter areas when a discernable degree of lift has taken place.

3 Make the application carefully so as to ensure that the colour is stripped evenly, and protect the root areas by inserting cotton wool strips between each mesh when working on middle lengths and/or ends.

4 When the required degree of lift has been achieved, rinse *very thoroughly* with warm water. The hair is now ready for tinting or bleaching.

BLEACHING

The hair's natural colour pigment is contained in the cortex, and consists of black/brown and red/yellow granules. The ratio of one pigment to another determines the colour of the individual hairs. The red/yellow pigment (phaeomelanin) will not oxidize as readily as the black/brown melanin, and for this reason, when a head of hair is bleached, we see the hair lightening through varying stages of red and yellow (see below).

Because of the resistance which this pigment has to being broken down, the development time can be quite prolonged, particularly where a high degree of lift is required on very dark hair.

HAIR COLOUR	BLEACHING STAGES
Dark brown to medium brown	Red
	Red/orange
Medium brown to dark blonde	Orange/red
	Orange
	Orange/yellow
	Yellow/orange
Hair lighter than dark blonde	Yellow
	Pale yellow
	Very pale yellow

If you understand the effects of bleach on hair you should be able to predict the amount of lift necessary for any tint application, whether it will be necessary to make more than one application and approximately how much time will be required for development. You will also be able to choose the correct type and strength of bleach.

TYPES OF BLEACH

It is important to use the *correct* type of bleach for the job in hand. Each has its own advantages.

Oil bleach This is a mild bleach for lightening the hair between $1\frac{1}{2}$ and 2 shades.

Emulsion (gel) bleach These are sometimes referred to as oil bleaches, but are capable of giving high degrees of lift, and have a controllable lightening action by adding activating powders. Emulsion bleaches are particularly recommended for whole head applications where subsequent toning is to be carried out.

Powder bleach This type of bleach is used for exceptional degrees of lift. Powder bleaches are ideal for use when highlighting or streaking, and on dark, resistant heads.

BLEACH APPLICATION PROCEDURES

Application of a bleach should be made carefully and cleanly.

Virgin head application

A full head bleach is generally followed by some form of colour application. It is necessary, therefore, to decide in advance what the final result is to be, so that the correct base is achieved. Choose the type and strength of bleach accordingly.

Method

1 Starting in the nape (as when tinting), apply the bleach to the middle lengths and ends of the hair first, leaving the first 25 mm (1 in) of the root hair untouched.

2 Place strips of cotton wool between each mesh to prevent the bleach from coming in contact with the roots.

3 Wait for a reasonable amount of lift before applying bleach to the roots. The reason for leaving the root area until the lengths and ends have lightened is that heat from the scalp will increase the rate of lift in this area.

4 When the required degree of lift has been achieved and the result is even, rinse the bleach away using warm water. Do not leave any trace of bleach in the hair.

5 Shampoo the hair using an antioxidant shampoo and towel dry. The intended colouring product can now be applied.

FOR A TARGET SHADE OF:	BLEACH TO:
Auburn/copper/red/mahogany	Orange
Light copper/warm chestnut	Orange/yellow
Soft ash/beige/light blonde	Pale yellow
Bleach toners	Very pale yellow

Regrowth method

1 Apply bleach to the root section only, being careful not to apply on the previously bleached areas.

2 Develop until the correct degree of lift has been achieved and remove as previously stated.

Temperature plays an important part in the speed of development of a bleach. When working in cold conditions or on a very resistant head, it may be advantageous to apply some form of external heat, either by use of a steamer or accelerator.

12

Hair styling

There are many techniques for cutting hair, and it is largely a matter of choice as to which method one adopts. Whichever method is chosen, one factor never changes, and that is the importance of creating a style for the individual.

Consideration must also be given to the overall suitability of a style, bearing in mind the following factors.

AGE OF CLIENT

It is wrong to create definite divisions between styles which are suitable for the younger client and styles for those of more mature years. Obviously, certain of the more outrageous modern styles would appear out of place on the older client, but the hairdresser should, within reason, show flexibility when discussing a hairstyle with the client. Many styles which at first may appear unsuitable, can be adapted very successfully to suit varying age groups.

FACIAL CHARACTERISTICS

The two most important facial characteristics to be considered are the nose and ears.

A large protruding nose will become even more pronounced if the hair is taken away from the face, and large ears or ears which stand out from the head need to be covered.

FACIAL SHAPE

By combing all of the hair away from the face, with the client sitting in front of a mirror, it is not difficult to observe the facial shape, which is of primary importance in creating a hair style suitable for the individual.

1 **Oval-shaped face** — (figure 12.1)

The oval-shaped face is considered to be the most suitable on which to create a hairstyle, because all hair lengths are accept-able. Short styles, or those of chin length, are exceptionally pleasing, and should follow the shape of the face.

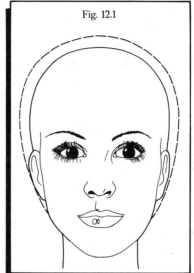

Fig. 12.1

2 **Pear-shaped face** — (figure 12.2)

An oval effect should be created by filling out the area above the chin. Do not build height on the top.

3 **Square-shaped face** — (figure 12.3)

Styles with soft lines should be selected, and it is impor-tant to create an illusion of roundness by emphasizing the hair at the temples and cheekbones. The facial shape is softened by creating a frame around it, covering the corners of the forehead.

4 **Triangular-shaped face** — (figure 12.4)

A pleasantly shaped face which is exceptionally well suited to asymmetric styles without fullness at the sides.

5 **Round-shaped face** — (figure 12.5)

An oval effect can be created by building height on the top. Full fringes should be avoided, as should width at the sides.

6 **Rectangular-shaped face** — (figure 12.6)

A more rounded look can be achieved by adding volume at the sides, and soft styles with fringes are recommended.

7 **Oblong-shaped face** — (figure 12.7)

Fringes will have a shortening effect on a long face, and fullness should be given to the sides. Loose, soft styles are flattering to an oblong-shaped face.

The styles illustrated are examples intended to show how a facial shape can be altered and flattered by giving consideration to its characteristics. By using some imagination, the talented stylist can create a number of different styles for the same client, all of which can be flattering.

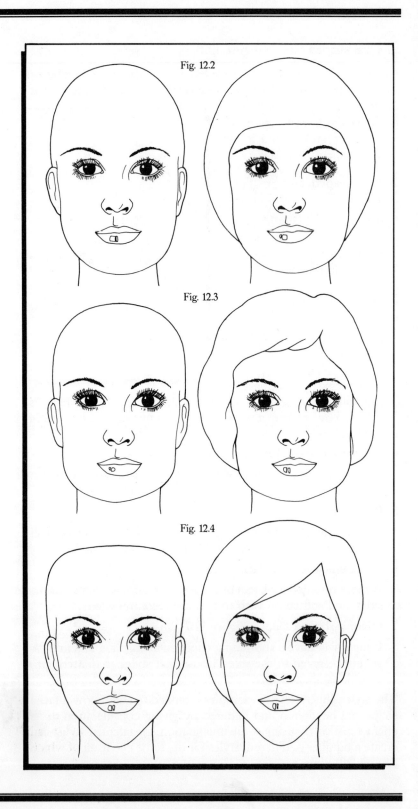

Fig. 12.2

Fig. 12.3

Fig. 12.4

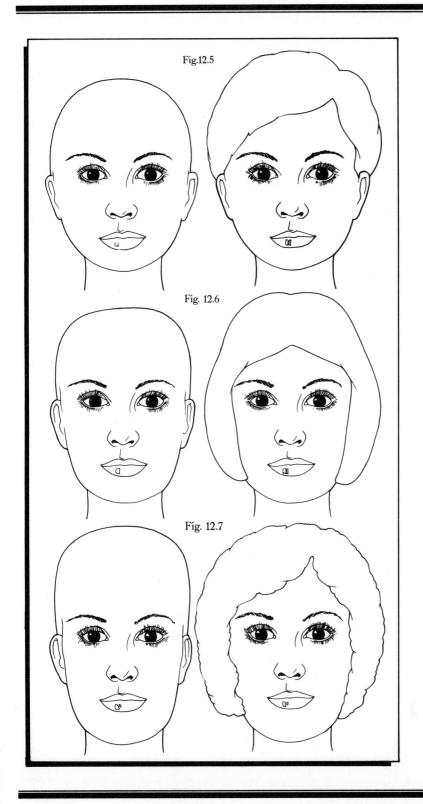

Fig.12.5

Fig. 12.6

Fig. 12.7

CUTTING THE HAIR

When it comes to learning how to cut hair, there can be no substitute for working, scissors in hand, under the guidance of an expert. However, certain principles do need to be understood before the first cut is made, and will be of great benefit in enabling the student to proceed confidently.

CUTTING PATTERNS

There are three principal cutting patterns in general use; the choice as to which one to adopt depends on the style required. Once learnt, these patterns can be varied and adapted to achieve many different styles.

1 **Basic layered cut** — (figure 12.8)

   The hair is cut to the same length all over the head, although it may be cut slightly shorter or longer around the hairline.

2 **Graduated cut** — (figure 12.9)

   Although this is a form of layering, the hair is cut in such a way so as to be shorter around the hairline, becoming gradually longer towards the crown.

3 **Reverse graduation** — (figure 12.10)

   The hair is cut so that each of the underneath sections is slightly shorter than the next when the hair is lying naturally.

Whichever principle is employed, it is still necessary to consider the shape line.

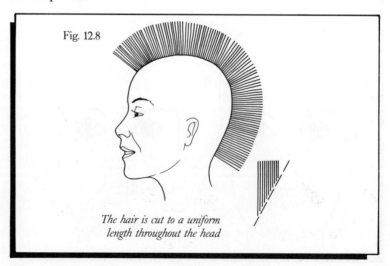

Fig. 12.8

The hair is cut to a uniform length throughout the head

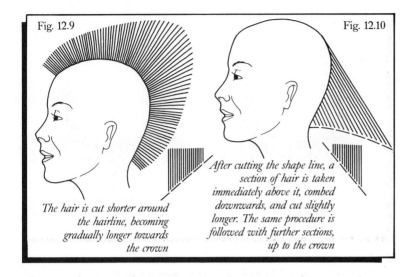

Fig. 12.9

Fig. 12.10

The hair is cut shorter around the hairline, becoming gradually longer towards the crown

After cutting the shape line, a section of hair is taken immediately above it, combed downwards, and cut slightly longer. The same procedure is followed with further sections, up to the crown

CUTTING THE SHAPE LINE — (figure 12.11 and figure 12.12)

Method

1 Take a parting down the centre of the head from the forehead to the nape (X–O–Y).

2 Take another parting from the crown to just in front of the ear on one side of the head (O–P).

3 A further parting is now taken between points X and P, leaving sufficient hair on which to cut the shape line. The rest of the hair is clipped up out of the way.

4 The same procedure is followed into the nape section, through points P to Y.

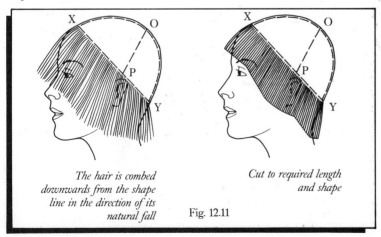

The hair is combed downwards from the shape line in the direction of its natural fall

Cut to required length and shape

Fig. 12.11

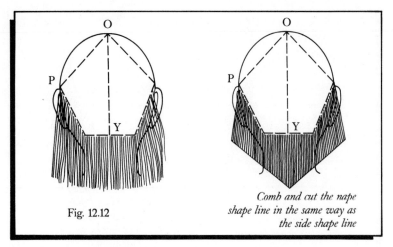

Fig. 12.12

Comb and cut the nape
shape line in the same way as
the side shape line

5 Section the other side of the head in exactly the same way. A continuous band of hair will now run around the entire head.

6 The hair is combed downwards and cut to the required length and shape.

COMPLETING THE CUT

Layered cut

After cutting the shape line, a vertical section of hair is taken at the crown of the head (figure 12.13 section 1) and held between the first two fingers of the hand, knuckles uppermost (figure 12.14). The width of this section should be between 6 and 12 mm ($\frac{1}{4}$ and $\frac{1}{2}$ in) depending on the thickness of the hair.

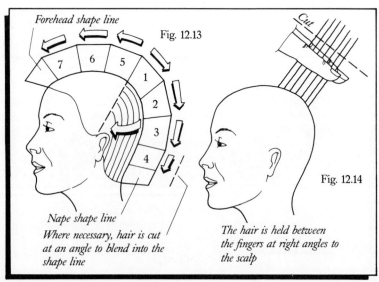

Forehead shape line

Fig. 12.13

Cut

7 6 5
1
2
3
4

Nape shape line
Where necessary, hair is cut
at an angle to blend into the
shape line

The hair is held between
the fingers at right angles to
the scalp

Fig. 12.14

The hair is held out from the head at right angles to the scalp and cut to the required length.

Continuing downwards towards the nape, section 2 is taken between the fingers, with the last quarter of section 1. This acts as a cutting guide to ensure that the hair is cut to the same length. Again, the hair is cut while being held at right angles to the scalp.

This procedure continues through sections 3 and 4 and into the nape's shape line. If the shape line is shorter or longer than the hair on the rest of the head, section 4 is cut at an angle, between section 3 and the shape line, to prevent a stepped result (figure 12.13). Once this central section from crown to nape has been completed, the remainder of the hair at the back of the head is cut in exactly the same way, in 6 to 12 mm ($\frac{1}{4}$ to $\frac{1}{2}$ in) sections, working outwards from the centre towards the ears. A section of the hair previously cut is taken between the fingers each time, so that it can be used as a cutting guide.

After completing the back of the head, sections 5, 6 and 7 are cut on the top of the head, between the crown and into the forehead shape line.

The remainder of the cut is carried out working across the head, from the top downwards into the side shape line, starting with section A at the crown (figure 12.15), and in the same way as when cutting the hair at the back of the head.

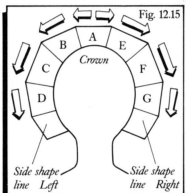

Fig. 12.15

Side shape line Left

Side shape line Right

Graduated cut

After cutting the shape line, a section of hair is taken right around the head, approximately 6 to 12 mm ($\frac{1}{4}$ to $\frac{1}{2}$ in) above the shape line, and the hair is combed downwards.

Starting at the centre of the nape, a horizontal section of this hair is taken between the fingers, along with a section of hair from the nape shape line to act as a cutting guide (figure 12.16).

The hair is combed downwards, at the same time slightly outwards from the head, and cut to the length presented by the shape line. When the hair is released from the fingers and combed downwards, a gentle graduation will be observed (figure 12.17). The higher the hair is lifted when cutting, the greater the degree of graduation will be (figure 12.18).

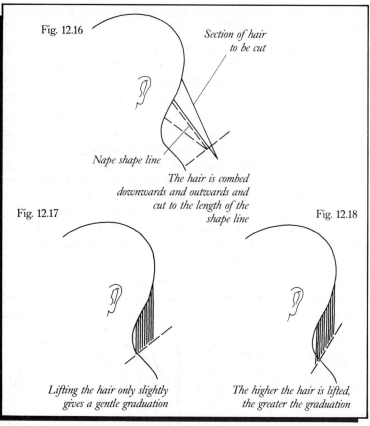

Fig. 12.16

Section of hair
to be cut

Nape shape line

The hair is combed
downwards and outwards and
cut to the length of the
shape line

Fig. 12.17

Lifting the hair only slightly
gives a gentle graduation

Fig. 12.18

The higher the hair is lifted,
the greater the graduation

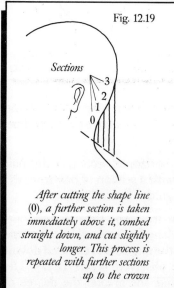

Fig. 12.19

Sections

3
2
1
0

After cutting the shape line
(0), a further section is taken
immediately above it, combed
straight down, and cut slightly
longer. This process is
repeated with further sections
up to the crown

Reverse graduation

This principle is employed when the hair is to turn under, as in the case of the pageboy style.

Sectioning is the same as when carrying out a graduated cut, working upwards from the shape line towards the crown, in 6 to 12 mm ($\frac{1}{4}$ to $\frac{1}{2}$ in) sections. Each section is combed straight down and cut slightly longer than the previous section underneath (figure 12.19). The hair must not be lifted away from the head as this will create a normal graduated effect.

BLOW-DRY STYLING

Although the shampoo and set is still a major salon service enjoying continued popularity, particularly with the older client, the majority of modern styles rely on freer movement and softer effects than are generally obtainable with the conventional roller pli. In this respect, the hand-held hairdryer adds another dimension to hairstyling techniques, although the principle employed is basically the same as for roller setting.

Both methods rely on the stretching of wet hair around a former, either a brush or roller, which is then dried in this new shape by applying a current of warm air. The advantage with blow-dry styling however, is that variable tension and movement can be obtained on each individual section of hair as it is being dried.

BASIC BLOW-DRYING PROCEDURE

Blow-drying a style will only be fully successful if the hair has been expertly cut and shaped beforehand. Indeed, the dryer and brush should be looked upon as tools for completing a style, not for creating one.

After washing the hair and cutting it if necessary, the hair should be moderately well towel-dried. If a blow-drying lotion is to be applied, the hair should be well towel-dried.

Comb through the hair to remove any tangles, and then comb into the shape of the intended style.

The dryer is generally held in the left hand, with the brush in the right hand. This is a matter of personal choice and it is a great advantage to be able to work with the dryer and brush in either hand.

The choice of brush is also a matter of personal preference. Radial (round) brushes of various sizes are ideal for turning the hair under, while the 'Denham'-type brush fulfils most of the requirements necessary for blow-drying.

Take a section of hair in the nape (figure 12.20, section 1), and clip the rest of the hair above it out of the way.

Turning the hair under

Starting at the centre of section 1, the brush is placed underneath a mesh of hair (figure 12.21), which shoud be no wider

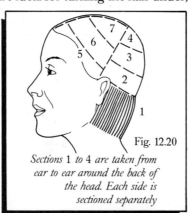

Fig. 12.20

Sections 1 to 4 are taken from ear to ear around the back of the head. Each side is sectioned separately

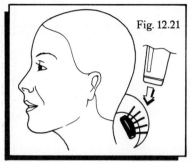

Fig. 12.21

than the usable area of brush, as it is important that the hair is held under tension throughout the drying period.

The airflow from the dryer is directed towards the points of the hair and drawn through from the roots to ensure that the entire mesh is uniformly dried. The brush is also drawn through the mesh from roots to points, slightly ahead of the dryer.

To obtain a very definite turn under at the ends of the hair, the brush is rotated in the hand, keeping the hair under tension at the same time as it must not be allowed to fly away from the brush. This skill may take a little time to master.

When this first mesh of hair has been completely dried, a further adjoining mesh is taken and dried in the same way. To avoid obvious gaps, include a small amount of the mesh of hair previously dried with the following mesh.

After completing section 1, continue working up the back of the head to the crown and then move to the sides (figure 12.20 sections 5,6,7).

Turning the hair out

The sectioning pattern is the same as previously described for turning the hair under.

Starting with a mesh of hair in the centre of the nape section, place the spines of the brush into the hair as shown in figure 12.22a, and draw it a little way down the mesh, at the same time lifting the hair slightly away from the scalp. Dry the root section thoroughly. The brush is then taken further down the mesh and turned so as to lift the ends of the hair (figure 12.22b). Continue drying, ensuring that the hair is always held firmly on the brush.

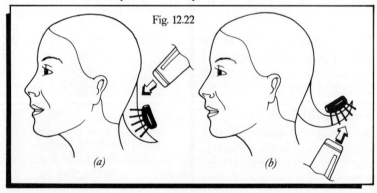

Fig. 12.22

(a)

(b)

When section 1 has been completed, proceed through sections 2, 3, and 4, or as far as is necessary to achieve the desired result.

Obtaining lift

To give the maximum amount of body to a style, a strong root lift is necessary. This is generally required on the top of the head, particularly at the crown.

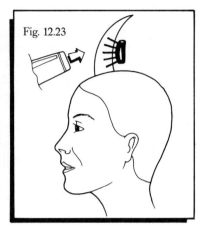

Fig. 12.23

To obtain this lift, the brush is pulled firmly into the root area of the mesh of hair to be dried, so that the hair is held at right angles to the scalp (figure 12.23).

The importance of correct airflow

It is imperative when styling with the hand-held dryer, that the flow of hot air travels from roots to points (figure 12.24), not points to roots (figure 12.25). There are two reasons for this. Firstly it allows complete control of the hair during drying, and secondly, it prevents damage to the cuticle layers.

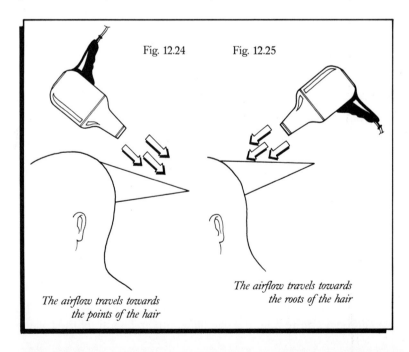

Fig. 12.24   Fig. 12.25

*The airflow travels towards
the roots of the hair*

*The airflow travels towards
the points of the hair*

ROLLER SETTING

Used with imagination and skill, rollers offer an excellent means of creating a wide variety of styles.

There is, however, more to learning how to use rollers correctly than is immediately apparent, and it should always be remembered that a first-class dressing can only be obtained from a first-class roller set. 'Acceptable' results can be obtained from poor rollering, but this does nothing for the professionalism and creative ability of the stylist, which the client has a right to expect. Covering the head with straight lines of rollers, with no consideration being given to the required finished style, may be quick and easy, but is bad practice and definitely not be encouraged.

Setting theory

When the hair is thoroughly wetted, the hydrogen and salt bonds (*not* the disulphide bonds) are broken and the hair becomes elastic. It is stretched around a roller and dried, the hydrogen and salt bonds reform, and the hair maintains its new shape until more water is absorbed into the hair, either from further wetting (immediate change of shape) or from the atmosphere (slow change of shape).

Although bonds within the hair are broken by wetting with water, the change which is brought about is a physical and not a chemical one. Before any physical change in the hair's structure takes place, the keratin is referred to as being in the *alpha* state (alpha keratin). Following any physical change, it is described as being in the *beta* state (beta keratin).

BASIC PRINCIPLES OF ROLLER SETTING

Choice of roller size

Two factors determine the choice of roller size: the degree of curl required for the style being created, and the length and nature of the hair being worked on. Smaller rollers will give a greater degree of curl, whereas larger rollers will give increased root lift, but softer wave movements.

Rollers should not be chosen on the principle that 'the smaller the roller, the longer the set will last', but should be selected for successfully achieving the desired style.

Correct sectioning

The size of section is dependent on the size of roller being used. It should be no more than the overall length and width of the roller, so that after the hair has been wound, the roller sits neatly on the section with no hair being pulled in from the sides. When taking the section, ensure that all partings are made cleanly. Maximum lift is obtained when the roller sits squarely on its section (figure 12.26), and this method is almost always used when setting the hair on the top of the head.

On areas of the head where less lift is required, generally at the back and sides, this can be achieved by dragging the hair as shown in figure 12.27.

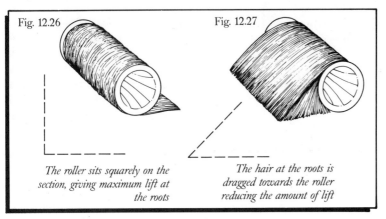

Fig. 12.26

Fig. 12.27

The roller sits squarely on the section, giving maximum lift at the roots

The hair at the roots is dragged towards the roller reducing the amount of lift

Winding the hair

Careful and accurate winding of the hair on to the roller is very important and should be practised extensively.

Fig. 12.28

Angle at which winding begins. The roller will sit squarely on the section

When the maximum amount of root lift is needed, the hair section is taken between the first two fingers of the left hand, as when cutting, and combed upwards and forwards of the section base (figure 12.28). The roller is placed into the mesh and drawn upwards until the points of the hair are resting on the top of the roller, and flat to it. Winding can now proceed, and great care must be taken in making sure that the points of the hair remain flat to the roller and are not trapped or twisted (this will result in 'fish hook' ends, which will be impossible to remove once the hair is dry).

When the roller has been wound and is sitting on its section, the securing pin is pushed through, so that the roller is held firmly, but not too tightly, as this will not only be painful for the client, but will also cause damage to the hair.

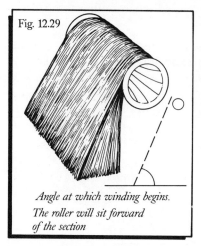

Fig. 12.29

Angle at which winding begins. The roller will sit forward of the section

When less root lift is required, the procedure is basically the same, except that the hair is combed slightly rearwards of the section base (figure 12.29).

When winding hair on to a roller, care must be taken to ensure that it is not bunched in the centre, but is evenly spread right across the roller. This is necessary if a smooth, well-finished style, is to be achieved.

Positioning of rollers

The direction and positioning of rollers on the head is determined by the movement required in the hair for the finished style.

Rollers should always be placed in the same direction as the hair is to be brushed when dressing out. Straight rows of rollers should be avoided (figure 12.30), as this will lead to gaps in the dressing which are difficult to remove without a considerable amount of backcombing

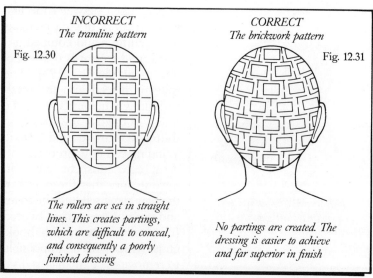

INCORRECT
The tramline pattern

Fig. 12.30

CORRECT
The brickwork pattern

Fig. 12.31

The rollers are set in straight lines. This creates partings, which are difficult to conceal, and consequently a poorly finished dressing

No partings are created. The dressing is easier to achieve and far superior in finish

The brickwork pattern (figure 12.31), allows for easy dressing and produces a style of superior finish.

Practice and experimentation with rollers is by far the best method for learning exactly what results can be achieved, and the enthusiastic student will be justly rewarded for his efforts.

PIN CURLING

Pin curling may not be practised as extensively as it once was, yet it has a number of useful applications, and is a skill which no hairdresser should be unable to perform.

Barrel curls

A section of hair is combed upwards from the head, in the same way as when preparing a section to accept a roller. The hair is held in the centre of the mesh, between the thumb and forefinger of one hand, and the points are curled round until they sit on the base of the section. A clip is used to secure the curl to the base of the section.

To prevent the curl from being flattened when covered by a net, a piece of cotton wool is inserted to act as a support.

The result obtained from a barrel curl is much the same as that obtained from a roller, but it is softer.

Flat pin curls

As their name suggests, flat pin curls lie flat to the head and are used to create definite wave movements, by a method of winding known as reverse curling. This is achieved by winding one row of pin curls in a clockwise direction, the following row in an anticlockwise direction.

When making a flat pin curl, the stem direction (direction of hair at the roots) is important in ensuring that a strong curl pattern is achieved from roots to points.

A curl which is clockwise in direction will start with the stem being pulled to the right. A curl which is anticlockwise in direction starts with the stem being pulled to the left.

The hair is then curled around the first finger of the right hand, (figure 12.32), and eased down on to the section, using the thumb and forefinger of the left hand, which also hold the curl in place until the clip has been inserted (figure 12.33).

When curling short hair, a tail comb can be used to act as a former, in place of the finger.

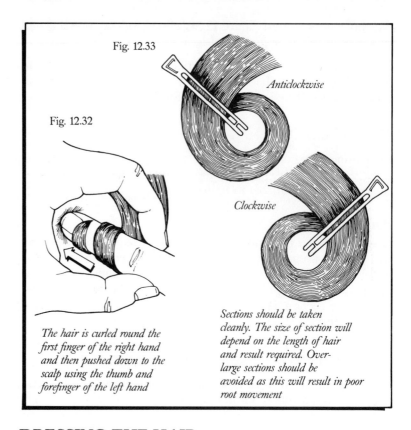

Fig. 12.33

Anticlockwise

Fig. 12.32

Clockwise

The hair is curled round the first finger of the right hand and then pushed down to the scalp using the thumb and forefinger of the left hand

Sections should be taken cleanly. The size of section will depend on the length of hair and result required. Over-large sections should be avoided as this will result in poor root movement

DRESSING THE HAIR

To create a first-class dressing, two objectives need to be achieved: firstly, a pleasing and well-balanced shape, and secondly, a neat, clean finish to the style, with movement and texture.

After removing all rollers and clips, the hair is brushed through very thoroughly and in all directions, until all traces of roller and pin curl sections have been removed. It is then brushed as closely as possible into the shape of the finished style.

If the hair has been set well, using the correct size of rollers, correctly positioned, a very satisfactory result can be obtained using only the brush, and many clients prefer this method of dressing, as opposed to the use of backbrushing or backcombing.

BACKBRUSHING

This is a technique which serves two very useful purposes. It gently binds the hairs together to give lift, at the same time maintaining wave and curl patterns, which can be used to good effect in giving movement and texture to the finished style. The hair is brushed through and placed as closely as possible into the shape and direction of the finished style. A section of hair is then taken and held between the first two fingers of the left hand. When working on the top of the head, the hair is lifted to an angle of approximately 45° from the scalp. When brushing out the back and sides of the head, the hair may be held straight out, or at varying smaller angles, depending on the amount of lift required. It must always be held in the direction of its final position.

The brush is held in the right hand and placed on top of the hair mesh, against the fingers. The edge of the brush closest to the fingers is lifted (figure 12.34), as the brush is made to travel through the length of the hair in an arc, by a gentle turning of the wrist.

It is pushed down the mesh towards the roots, lifting some of the surface hairs as it progresses (figure 12.35). Only light pressure must be used. The brush is not buried deeply into the mesh of hair, but is made to glide along the surface layers only.

The stroke is completed when the brush has travelled the full length of the mesh (figure 12.36), and is repeated until all the hair has been drawn from the fingers.

The style is completed by lightly brushing or combing over the surface of the hair, or, for a softer and more casual result, using the fingers and a four-pronged styling comb.

Fig. 12.34   *Angle of brush on entry*

Fig. 12.35   *Angle of brush through the stroke*

Fig. 12.36   *Angle of brush on completion of stroke*

Hands are a valuable aid in the formation of a hairstyle, and can be used to press the hair into shape, and flatten areas of too much height. Another invaluable aid is the mirror, which should be used constantly during dressing out, to ensure that the corect height, shape and balance are being achieved.

When backbrushing the hair, it is a good idea to start at the front of the head, working backwards towards the nape and downwards into the sides, so that each mesh of hair is brushed into the section previously backbrushed.

BACKCOMBING

The backcombing principle is basically the same as that of backbrushing, but with this method of dressing, the binding of the hair is greater. Backcombing is usually employed where a style requires the maximum amount of height.

Sections of hair are taken in the same way as for backbrushing, and held either at right angles from the head, where maximum lift is needed, or at angles closer to the head where less lift is necessary.

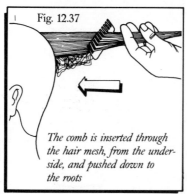

Fig. 12.37

The comb is inserted through the hair mesh, from the underside, and pushed down to the roots

The comb is inserted right through the hair mesh on the underside, and close to the roots. Only the fine teeth are used, therefore the hair mesh should be no wider than this section of the comb. The comb is pushed towards the head, binding some of the hairs together, and causing the hair to stand upright at the roots (figure 12.37).

Each consecutive stroke starts further along the mesh towards the points of the hair, although only the highest of dressings will necessitate backcombing the entire mesh.

After backcombing, the hair can be smoothed over using either a brush or the large-toothed end of a comb.

Dressing the hair, as with any other aspect of styling, will take time to master. Even when basic techniques have been learnt, it will be found that they will need to be varied when used on heads of different hair textures and curl patterns.

Frustration is something which the hairdressing student has to accept – perseverance is essential and will be well rewarded.

13

Setting and blow-drying lotions

SETTING LOTIONS

Most modern setting aids are lotions which have a base of synthetic resins such as polyvinylpyrrolidone.

The purpose of a good setting lotion is to prolong the life of a set, make the hair more manageable when brushing out, stop the hair becoming flyaway, and give the hair a shine.

There are two main factors which govern how long a set will last, although they are directly linked. Firstly, the natural 'pull' of the hair, and secondly, the hair's ability to absorb moisture from the atmosphere.

Hair has a natural shape to which it will always return: straight, wavy, or curly. This process is accelerated by the rapidity with which moisture is absorbed into the hair. A setting lotion places a protective coating around the cuticle which helps to hold the hair in the shape in which it has been set, at the same time creating a shield against moisture. It is often believed that the stronger and crisper the setting lotion, the longer the set will last. This is not necessarily the case, and the choice of setting lotion should be dependent on the type of hair to which it is to be applied.

CHOICE OF SETTING LOTION

Setting lotions fall into two categories: basic and dual-purpose.

Basic setting lotions are of two main types: crisp-result lotions and soft-result lotions. Their function is the same but each is better suited to particular hair types. The softer setting lotions are particularly suitable for fine to medium textured hair, and hair which has been bleached, tinted, or permanently waved.

Crisp-result lotions are more suitable on strong-textured hair, and hair which is lank and difficult to manage.

A dual-purpose setting aid does two jobs at once. Apart from increasing the life of the set, it has a secondary function of doing one of the following things: imparting delicate colour tones, conditioning the hair and reducing the flow of sebum along the hair shaft, thus preventing the hair from becoming greasy so quickly.

It is obvious, from the information given, that before any setting aid is recommended to a client, the following factors need to be considered:

1 Hair texture.
2 Whether the hair is bleached or tinted.
3 Whether the hair is permanently waved.
4 Is the hair lank and lifeless?
5 Is the hair greasy or dry?

APPLYING A SETTING LOTION

After the hair has been shampooed, it should be well towel-dried, and the setting lotion applied in drops, evenly throughout the head. Comb through thoroughly and proceed with setting. If the correct procedure of application is not followed, the hair may become sticky when dry, and probably covered with white specks.

A setting aid is beneficial to every client.

BLOW-DRYING LOTIONS

The functions fulfilled by a good blow-drying lotion are:

1 Reduced flyaway tendencies caused by static electricity.
2 Prolonged life of a style.
3 Easier blow-drying, by allowing the brush to pass through the hair more easily.
4 Added shine.

Blow-drying lotions are applied in the same way as setting lotions, to shampooed, towel-dried hair.

14

Care of wigs and hairpieces

The majority of modern wigs are machine made and manufactured from synthetic fibres. Hand knotted, human-hair wigs, are however still in demand, although the art of producing them has almost vanished from the hairdresser to small, specialist firms who manufacture high-quality wigs.

Although the hairdresser may no longer produce the wigs, many are brought to the salon for cleaning, colouring, and even permanent waving.

CLEANING METHODS

Washing
Although the washing of wigs and hairpieces is not particularly recommended, the wire-base and bonded-base types can be shampooed provided very great care is taken.

Method
1 As wigs and hairpieces have no natural source of oil, use a shampoo for naturally dry hair.

2 Hold the base of the wig or hairpiece horizontally so that the hair hangs straight down. It is a good idea to obtain the assistance of another staff member for this purpose.

3 Soak the hair with warm water, being careful not to wet the base any more than is necessary.

4 Apply the shampoo and work in gently. Be careful not to rub too hard as this will create tangles, and any subsequent combing will damage the foundation, causing some of the hair to be pulled out.

5 Rinse thoroughly. To remove excess moisture, press the hair gently in the folds of a towel.

6 Apply a conditioning rinse, comb through and rinse well.

7 Proceed with styling.

'Dry' cleaning

This method of cleaning must be used on all wigs and hairpieces which have knotted foundations, and on all other hairwork wherever possible. A solvent is used in this process, the most successful being a mixture of carbon tetrachloride and petroleum spirit. Carbon tetrachloride can also be used on its own.

Method

1 Pour the solution into a dish. This must *not* be done in a confined space, as the fumes are very dangerous. Ensure that there is adequate ventilation and do not hold your head too close to the solution.

2 Brush through the wig or hairpiece thoroughly, and gently work it up and down in the solution with the base uppermost. Do not rub the hair but gently press it from time to time, until all of the dirt is removed.

3 Pour out a fresh amount of cleaning solution and repeat the process.

When any hairwork has been dry cleaned, it should be allowed to dry completely before applying any other product.

COLOURING A WIG OR HAIRPIECE

When a client first purchases a wig or hairpiece, she chooses one which matches her own natural colour or the colour to which her hair is tinted. If she wishes to change the colour of her own hair, the chances are that she will also wish the colour of her wig or hairpiece to be changed. In all cases where these are manufactured from synthetic fibres, re-colouring should *not* be attempted. The client should be advised to purchase a new wig or hairpiece to match the new hair colour.

Good quality human hair wigs and hairpieces can in many cases be coloured, but extreme care must be exercised.

Strand tests can give an indication as to the degree of success which can be achieved, but should not be taken as a guarantee.

Using temporary colours

Where only a slight change in colour tone is required, temporary colours such as coloured setting lotions can be used successfully.

The application is simple. Apply the product to the hair after first dampening with water, again ensuring that it is kept away from the base. Setting or blow-drying can then be carried out in the normal way.

Using semi-permanent colours

A great advantage when using semi-permanent colours on wigs or hairpieces is that unlike temporary colours they do not require the addition of an oxidant, which could be damaging to the hairwork base. The colour range is wide, although a semi-permanent colour cannot be used to lighten the hair. It can, however, darken the hair considerably and add a wide choice of tones.

Application is simple but requires a great deal of care. Apply the colour with a brush or sponge, depending on the product, to within 12 mm ($\frac{1}{2}$ in) of the base. If the hairpiece or wig is resting on a flat surface or mounted on a block, strips of cotton wool should be placed between each mesh at the base (as the colour is applied), to prevent the preparation reaching the base. After the appropriate processing time, rinse thoroughly and carefully, and blot with a towel to remove excess moisture.

Using permanent colours

With a permanent colour, the choice of shade is extensive, and there is the additional advantage of being able to lighten the hair. Extreme caution must be exercised, however, when using this form of colourant. As previously stated, oxidants are very damaging to hairwork bases. Application is made in the same way as it would be to a natural head of hair, but to within 12 mm ($\frac{1}{2}$ in) of the base. Cotton wool strips should be placed between each mesh to ensure complete protection. The procedure for removal is as described in the section on washing, preferably using an antioxidant shampoo.

MEASURING A CLIENT FOR A WIG

Because of the excellent range of manufactured wigs available, it is seldom necessary to measure a client for a wig. On occasions, a client who demands only the best, will insist on a wig being made exclusively for her. The cost of a wig made in this way is usually high.

The measurements required by a wigmaker are:

The circumference — (figure 14.1)

From the forehead to the nape — (figure 14.2)

Ear to ear around the front — (figure 14.3)

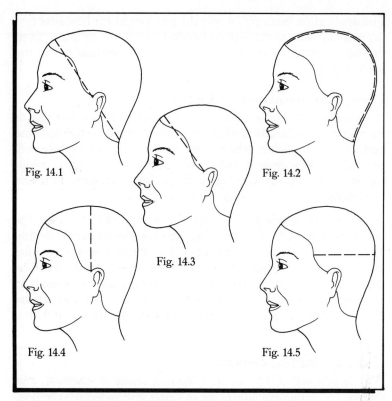

Fig. 14.1

Fig. 14.2

Fig. 14.3

Fig. 14.4

Fig. 14.5

Ear to ear over the top of the head — (figure 14.4)

From temple to temple around the back of the head — (figure 14.5)

The width of the neck should also be measured and it should be stated whether the wig is for a man or woman, and if a parting is required, where this is to be.

15

Aspects of barbering

If you walked down any high street before the 1960s you would have seen at least two shop frontages with red and white poles attached to them. These poles signified the trade of the barber. Today however, there are few of these shops left and even the name 'barber' has largely disappeared — to be replaced by the more modern image created by the title 'Gentlemen's hairstylist'.

It was the cultural changes brought about by the young people of the 1950s which over the next 15 years or so were to have such a major influence on the services offered by the barber.

Rock and roll not only gave teenagers of the day a music of their own, it also brought with it different attitudes towards fashion, and in particular, hairstyles. Short back and sides were not for the new generation of young men, although many barbers thought these longer hairstyles to be no more than a passing fad that would inevitably disappear. This was not to be the case, and many of those barbers who firmly refused to adapt to the new trends were themselves the ones to disappear.

A demand does of course still exist for the traditional barber, who now, more than ever, needs to be a skilled technician.

Many of the services offered in the ladies' salon have now found their way into the gentlemen's salon, for example, permanent waving and colouring. Certain services however, still remain firmly in the domain of the barber and are the ones dealt with in this section.

SHAVING

Commonly referred to as the cutthroat, the straight or open razor (figure 15.1), needs considerable skill if it is to be used safely and effectively. It is not a piece of equipment to be used or even handled without the utmost care and attention.

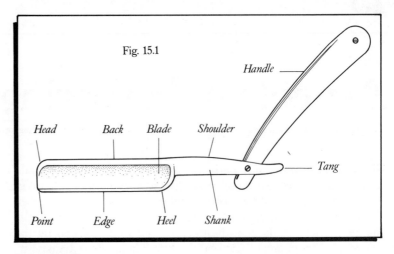

Fig. 15.1

Head   Back   Blade   Shoulder   Handle   Tang

Point   Edge   Heel   Shank

RAZORS

The straight or open razor has evolved over many centuries. There is evidence to show that primitive man honed flint to produce a crude hair-cutting implement.

The modern razor is manufactured from very high-quality steel, ground to perfect sharpness, and encased in a handle of wood or plastic. Two types of razor are in common use: the hollow-ground (the most recent development) and the French or solid razor. They differ in that the blade of the hollow-ground razor is concave in cross-section (figure 15.2), whereas the French razor has perfectly straight sides (figure 15.3). Although the latter is still in use, the hollow-ground razor has the advantage of being lighter in weight, and has a longer lasting and keener blade. Softer steel is used in the manufacture of the French razor. Consequently, it loses its keen cutting edge quite quickly and more skill is required in sharpening this type of razor. The stropping technique is more difficult to master, but it does have the advantage over the hollow-ground razor in that it seldom needs setting on an oilstone.

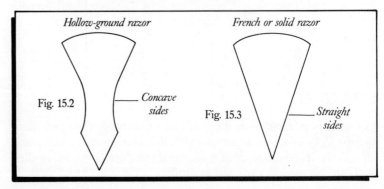

Hollow-ground razor

French or solid razor

Fig. 15.2   Concave sides

Fig. 15.3   Straight sides

Three major factors need to be considered when choosing a razor. They are balance, temper and grind.

1 **Balance**

This refers to the comparative weight of the blade to that of the handle. Ideally, they should be the same.

2 **Temper**

As a result of special heat treatment during manufacture, steel can be made harder or softer. The correct degree of hardness is essential if the best cutting edge is to be ensured. Hollow-ground razors are generally made from medium-tempered steel, whereas softer steel is used for the French razor.

3 **Grind**

This is the shape of the blade when viewed in cross-section. Between the shoulder and the bevel of the hollow-ground razor, a distinct hollowness can be seen. The French razor has straight sides.

KEEPING A RAZOR SHARP

A razor has an exceptionally fine cutting edge which must be maintained. To do this, a piece of equipment known as a strop is employed, of which there are two types. The type of strop required will depend on the razor to be sharpened: for the hollow-ground razor – the hanging strop (figure 15.4); for the French razor – the French or solid strop (figure 15.5).

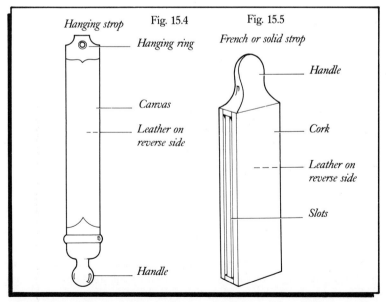

Hanging strop   Fig. 15.4   Fig. 15.5

Hanging ring

French or solid strop

Handle

Canvas

Leather on reverse side

Cork

Leather on reverse side

Slots

Handle

The hanging strop

Ideally around 600 mm (24 in) in length and 62 mm (2½ in) wide, the hanging strop consists of a strip of leather on one side, and a strip of canvas on the other, stitched together at each end. At one end is a handle, and at the other end, a metal ring for attaching the strop to a suitably placed hook, a little above waist height. The strop is prepared for use by applying tallow or fine oil to the leather side, which is allowed to soak in. To prepare the canvas side, the strop is attached to the hook and is firmly stretched. Hard soap is applied and rubbed in using the tips of the fingers. An even, smooth surface can best be achieved by rubbing over the canvas with a round glass bottle. From time to time, the strop should be cleaned, followed by this same procedure. This will ensure many years of good service.

The French or solid strop

As its name suggests, this strop is solid, being manufactured from a piece of wood approximately 300 mm (12 in) in length. One side is covered in leather and the other side in cork. Two special pastes are required for preparing the French strop, a coarse-grade paste for the cork side, and a softer-grade paste for the leather side which is used for the final finishing and polishing of the razor. Only small amounts of paste should be applied, the procedure being the same as for the hanging strop. Some French strops have slots cut through the sides, which gives resilience to the stropping area.

RAZOR STROPPING

For hollow-ground razors, the hanging strop is used.

Method

1 Attach the strop to the hook, canvas side uppermost.

2 The handle of the strop is held firmly in the left hand. Keeping the strop in a horizontal position, the operator pulls it towards him, so that a firm but not over-strong pressure is exerted.

3 The razor is opened out straight and held in the right hand with the thumb over the tang and the other fingers curled firmly, but not too rigidly, underneath the handle.

4 The razor is placed across the strop at the end closest to the operator, with the edge of the blade facing towards him.

5 Ensuring that the razor is lying perfectly flat on the strop, it is pushed away from the operator until it has covered about two-thirds the length of the strop. The razor must be kept perfectly flat throughout the length of the stroke.

6 Without lifting the razor from the strop, it is turned over on its back on to its other side, and then with the blade facing away from the operator, drawn towards him. This procedure is usually carried out until each side of the razor has received six strokes (more will be required for a very blunt razor) and then the strop is turned over so that the leather side can be used.

7 In the same way as previously described, the razor is moved up and down the strop to give a keenness and polish to the edge.

The solid strop is required for sharpening the French razor.

Method

1 Hold the handle of the strop firmly in the left hand, the cork side uppermost, and place the end on a solid surface so that it rests against a stop, to prevent it from slipping. The strop should be lifted by the handle to approximately 25° from the horizontal.

2 The first stroke is made by pushing the razor away from the operator, the second stroke is made towards him.

3 Unless the razor is very blunt, continue stropping until the razor has received six strokes on either side.

4 Turn the strop over so that the leather side is facing uppermost and give each side of the razor an even number of strokes. This will give the razor a final finish and polish.

SETTING A RAZOR

From time to time, it will be necessary to put a new edge on the razor. This will become apparent when stropping fails to give the required degree of sharpness.

To do this on a hollow-ground razor, a special type of sharpening stone is needed, the most popular being the Belgian hone. This is a yellowish stone, about 225 by 50 mm (9 x 2 in), cemented on to a slate base. When working on a French razor, a smaller, fine grain Belgian hone is used, about 100 by 50 mm (4 x 2 in); apart from the fact that the strokes are shorter and crisper, less pressure is exerted, and the edge is raised slightly during the setting procedure, the method is the same as that which is applied to the hollow-ground razor.

The stroke patterns are illustrated in figures 15.6 and 15.7.

Method

1 The hone should be thoroughly cleaned and then lubricated, using oil, water, or soap lather. The thinner the lubricant, the coarser the cutting action will be.

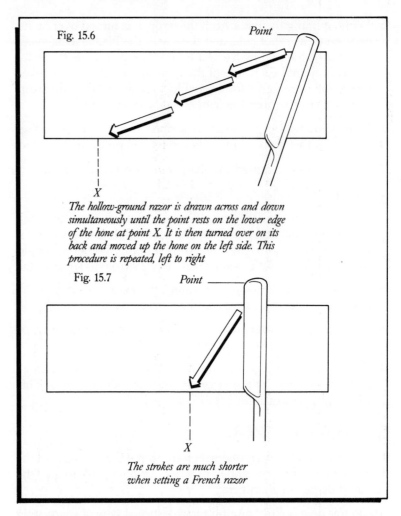

Fig. 15.6

Point

X

The hollow-ground razor is drawn across and down simultaneously until the point rests on the lower edge of the hone at point X. It is then turned over on its back and moved up the hone on the left side. This procedure is repeated, left to right

Fig. 15.7

Point

X

The strokes are much shorter when setting a French razor

2 The operator should be seated in front of a work surface of suitable height, with the hone placed to the right-hand side.

3 Holding the razor in the right hand, the blade is placed on the right hand side of the hone, at an angle of 45°, ensuring that the blade is perfectly flat. The edge of the blade should be facing towards the centre of the hone.

4 The razor is now drawn edge first towards the middle of the hone, at the same time being drawn downwards. At the end of the stroke, the point of the razor should be resting on the lower edge of the hone.

5 The razor is turned over on to its other side and slid upwards so that it is again at an angle of 45°, but on the left-hand side of the hone. The second stroke is now made from left to right.

6 Keeping the razor perfectly flat, these strokes are repeated, ensuring that each side receives an equal number of strokes.

7 When a suitable edge has been achieved, the razor should be carefully wiped clean. It is now ready for stropping and testing.

RAZOR TESTING

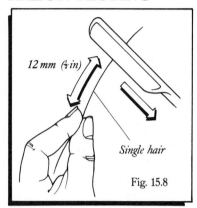

12 mm (½ in)

Single hair

Fig. 15.8

From the point of view of safety, the best method for testing a razor is to take one single hair, hold it between the thumb and forefinger, leaving about 12 mm (½ in) of it protruding, and draw the razor across it (figure 15.8). If the hair cuts cleanly, no further setting is required.

SHAVING PROCEDURE

However enthusiastic one may be, it is not recommended that any trainee be allowed to use a razor on an actual face without some prior practice. It takes time to become proficient at handling a sharp instrument against such a delicate surface. The best object for this practice is a balloon, blown up reasonably hard and to which lather can be applied. This makes an ideal training aid on which to learn the correct strokes and to obtain a 'feel' for the razor.

One point which must be borne in mind, is that every stroke of the razor not only removes hair from the face but also a portion of the skin's outer layer. The number of strokes taken therefore, should be as few as possible to achieve a first-class shave.

Method

1 Using capes and towels, ensure that the client is fully protected and reclined in the chair at an angle suitable for easy working.

2 Fill the shaving mug with very hot water (this helps to soften the beard) and, using a good-quality shaving brush and soap, generate a rich lather.

3 Using the brush in circular movements, apply to the entire area to be shaved, being careful not to allow soap to enter the eyes, nostrils or mouth.

4 Hold the razor as shown in figures 15.9 and 15.10.

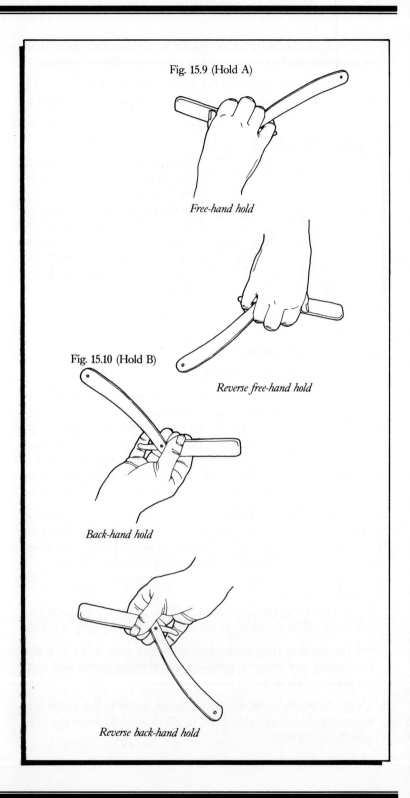

Fig. 15.9 (Hold A)

Free-hand hold

Reverse free-hand hold

Fig. 15.10 (Hold B)

Back-hand hold

Reverse back-hand hold

5 Standing on the client's right-hand side, place the forefinger of the left hand on the skin, behind the first razor position and push the skin upwards so that it becomes stretched.

6 With the razor almost flat to the face, proceed to draw it across the skin, the angle and direction of the strokes being as in figure 15.11. The movement of the blade must be smooth and free from jerks. The length of each stroke will vary depending on which part of the face is being shaved, but no stroke should exceed 75 mm (3 in).

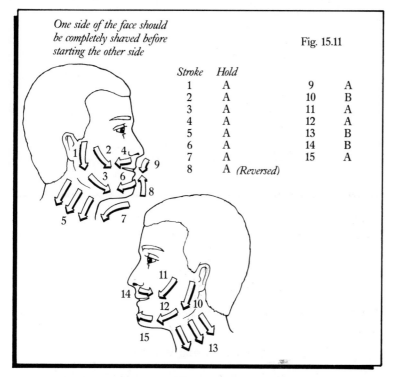

One side of the face should be completely shaved before starting the other side

Fig. 15.11

Stroke	Hold		
1	A	9	A
2	A	10	B
3	A	11	A
4	A	12	A
5	A	13	B
6	A	14	B
7	A	15	A
8	A (Reversed)		

7 Complete the right-hand side of the face before moving to the left-hand side, stretching the skin each time a stroke is made.

8 Extra care must be taken in the area of the chin. Use the centre of the blade only, slightly lifting the back of the razor. New strokes should not begin on prominent parts of the face such as the chin. These areas should be covered by the continuation of a stroke from a flatter area.

9 The first time 'overshave' is now completed.

10 Apply further lather and proceed with the second time 'overshave', the strokes this time being against the grain, as illustrated in figure 15.12.

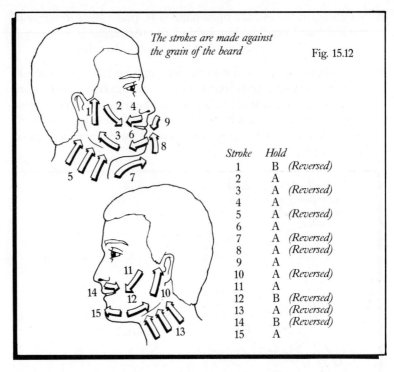

The strokes are made against the grain of the beard

Fig. 15.12

Stroke	Hold	
1	B	*(Reversed)*
2	A	
3	A	*(Reversed)*
4	A	
5	A	*(Reversed)*
6	A	
7	A	*(Reversed)*
8	A	*(Reversed)*
9	A	
10	A	*(Reversed)*
11	A	
12	B	*(Reversed)*
13	A	*(Reversed)*
14	B	*(Reversed)*
15	A	

11 On completion, sponge the face with clean water, dry gently with a face towel and apply aftershave lotion. A hot towel treatment is sometimes given after shaving.

HOT TOWEL TREATMENTS

A hot towel treatment can be carried out on the face for a number of reasons:

1 As an invigorating and cleansing treatment.

2 To prepare the face before shaving, or in between the first and second time overshave to soften a strong beard.

3 Following shaving. This helps to remove any traces of soap left in the skin pores and leaves the skin feeling fresh and clean.

Method

1 Two white face towels are required. As one towel is removed from the face, a second must be ready for immediate application.

2 Each towel is folded in half lengthways and then each end is brought into the middle to enable the towel to fit into the steamer cabinet. Where no towel steamer is available, do not make these folds into the centre; instead, the towels (folded down the centre only) are placed in a bowl of very hot water, the ends being left to hang over the sides.

3 The client is suitably protected and seated in a reclining chair, his head in the same position as for shaving. The top of the head should be covered with a dry towel for protection.

4 Remove one hot towel from the steamer. If using hot water, take the ends of one towel and twist in opposite directions to remove as much water as possible. In either case, the towels should then be opened out (though still folded down the centre) and held underneath, on the palms of the hands, each hand being one third of the way in from each end.

5 As quickly as possible, the towel should be applied to the client. The operator stands behind the client and places the centre of the towel under the chin. Both ends are brought up to the top of the head and crossed over, ensuring that the entire face is covered, with the exception of the nose, to allow for easy breathing (figure 15.13).

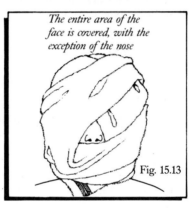

The entire area of the face is covered, with the exception of the nose

Fig. 15.13

6 The first towel is left for 2 minutes and then replaced immediately with the second towel.

7 Following the second application, the face should be thoroughly towel dried and aftershave lotion applied. This final procedure is not used, of course, when shaving is to follow.

FACIAL MASSAGE

The purpose of massage is to keep the skin soft and supple and to improve circulation.

Massage of the face can be carried out in two ways: by hand and by use of the vibro massager.

In hand massage, three different techniques are employed, these being effleurage, petrissage and tapotement, which are carried out in the order listed.

1 Effleurage

For this form of massage, which is designed to soothe, the fingers and palms of the hands are used, with only very light pressure.

A gentle, stroking movement, is applied to the forehead, around the eyes, to the sides of the nose and under the chin on the neck.

2 Petrissage

The tips of the fingers are used to produce a kneading movement on the fleshy areas of the face. Using light pressure, the flesh is taken between the thumb and fingers and gently pinched.

3 Tapotement

The fingertips are used in a light tapping action on the bony areas of the face, such as the jawbones and forehead. The fleshy parts, such as the cheeks and under the neck, undergo a gentle slapping action using the palms of the hands. Rapid movement is necessary and although only light pressure is used, it should be sufficient to make the skin tingle.

The preparation of the client is the same as for shaving, and a towel is placed over the head to protect the hair from the massage cream.

Method

1 Make one hot towel application and ensure that the face is clean and free from grease.

2 Apply the appropriate massage cream to the face. For a dry skin use cold cream, for a normal or slightly greasy skin use lemon cream, and use rolling cream for a very oily skin.

3 Apply a second hot towel to assist absorption of the cream into the skin pores.

4 Remove the hot towel after 2 minutes and if necessary apply more cream.

5 Carry out all of the massage movements as previously described, for a period of 4 minutes each.

6 Apply a further hot towel and on removal thoroughly wipe the face, ensuring that no stickiness remains.

7 Apply a spirit lotion or witch hazel, to close the skin pores.

8 Lightly dust the face with talcum powder and massage in.

USING A VIBRO MASSAGER

When using the electrical vibro massager, the initial preparation of the client (hot towels and massage cream application) is exactly the same as for a hand massage.

Two attachments are used: a rubber (or plastic) suction cup and a sponge attachment.

Method

1 The suction cup is fitted to the vibro machine and given a light coating of massage cream.

2 Starting at the centre of the forehead, begin the massage by moving the vibro in small circular movements, working down one side of the face, over the cheek, along the jawbone, ending up under the neck at the point of the chin.

3 Continue from this point, working up the other side of the face, until the starting point on the forehead is reached.

4 Repeat the operation, this time in reverse, so that the side of the face last treated, is now treated first.

5 Place the left hand over each eye in turn and place the suction cup on to the hand. Do not apply massage directly to the areas of the eyes.

6 Place the index finger of the left hand on each side of the nostril in turn and apply massage to the other side.

7 On completion, a third hot towel is applied, the face blotted with a dry towel and spirit lotion is applied.

8 Lightly apply talcum powder and, using the sponge attachement on the vibro, massage into the face in circular movements.

A diagram of the vibro massager is show on p. 23, figure 5.1.

FRICTION LOTIONS

Friction lotions are used to give an invigorating and fresh feel to the scalp, although it must be stressed that they have *no* cleansing action. They are pure spirit lotions.

Method

1 Shampoo the hair in the usual manner and thoroughly towel dry.

2 Transfer the client from the shampoo bay to the chair, and ensure that he is adequately protected by towels from any lotion which might drip.

3 Use a gentle fingertip massage and at the same time apply the friction lotion over the entire area of the scalp.

4 Stand behind the client, and again using the fingertips, but with a firmer pressure, give the scalp a vigorous massage for about 2 minutes. Work in a set pattern to ensure that no part of the scalp is missed.

5 It will be noticed that on completing the massage, most of the lotion will have evaporated, but a towel should be used to ensure that the hair is completely free of lotion before styling with the blow-dryer begins.

TRIMMING BEARDS AND MOUSTACHES

When trimming a beard, consideration must be given to the shape of the face as it is when choosing a hairstyle.

A long, narrow face, with sunken cheeks, should be trimmed so that as much fullness as possible is given to the sides. The hair should be trimmed relatively short under the chin.

A full, round face, can be made to appear longer by closely cropping the sides and cutting to a point at the chin. Underneath, the neck should be closely shaved.

When working on a face with a long chin, cut the hair square across and under the chin, blending it into the sides.

Method

1 Prepare the client as for shaving, and seat him in a suitably reclined chair. To prevent small hair clippings from entering the eyes, moisten two cotton wool pads with water, and place them over the eyes.

2 Remove all tangles from the beard, using a brush followed by a comb, and arrange as near as possible into the final desired shape.

3 Using the scissors over a comb, begin cutting on one side of the face, working downwards from the ear. A clearly defined shape can be achieved later using the electric clippers.

4 Complete cutting on both sides of the face, and then proceed to cut under the chin. Stand back from the client from time to time, looking to make sure that a well-balanced shape is being achieved.

5 Finally, using free-hand cutting, go over the beard once more, again checking that the final result is evenly balanced.

6 If a clean-shaven area is required under the chin, use the electric clippers.

7 Clippers are also used for trimming the moustache, to give a clean line along the edge of the top lip, and to remove any straggly hairs. To cut the moustache to a short length, cut over the comb, using the points of the scissors.

16

Caring for the client

The client may not always be right, but she is the most important person in the salon. Indeed, without clients there is no salon. However skilled a stylist is and whatever technical knowledge he possesses, these attributes are wasted if there is no genuine desire to please the client and make her visit an enjoyable experience. In fact, this philosophy of caring is the responsibility of every member of staff from apprentice to manager.

Interest in the client can be shown in many different ways, ranging from using only the best quality products to providing clean, comfortable surroundings; from offering a high standard of service to a pleasant and sociable atmosphere.

SELL YOURSELF

The first thing a company sales representative learns is that when calling on a prospective customer, he should sell himself first and the product second. In other words, if he creates a favourable impression on the customer by his attitude and approach, he not only has a good chance of achieving an order, but is also creating in the mind of the customer a good impression of the company by which he is employed. All salon staff members should sell themselves to the client in the same way, and with sincerity.

Unlike the situation facing the average salesman, where he must convince the potential customer of the benefits of his products over others, and surmount the barriers they set up against him when it comes to parting with money, salon clients actually make appointments to spend money. This situation is all too often taken for granted and leads to the wrong mental attitude of some hairdressers in relation to promoting themselves and their salon in general.

The client may have made an appointment to spend money at your salon this week, but the question is, will she do the same thing next week or will her name be on the appointment book of the salon down the road?

THE APPOINTMENT

The first contact the client has with the salon is when she makes her appointment. There are two ways for her to do this: firstly by telephoning, and secondly by visiting the salon in person. If this first encounter is favourable, the client will build a good mental impression of the business. If it is not favourable, she will paint a poor mental picture and may not even keep her appointment. Any staff member responsible for taking appointments should realise how important this area of salon routine is in the overall success of the business.

When answering the telephone, speak in a clear, friendly and positive manner. The client should not be left to seek information when making her appointment, but instead, the questions should be put to her. For example, on what day and at what time would she like her appointment? Has she visited the salon before, and does she require the services of any particular stylist? This is just as important with a first-time client, as she may be booking her appointment on the recommendation of an already regular client.

What services does she require? If a particular stylist is asked for but is unavailable owing to prior bookings, an alternative time or different stylist should be recommended, with confidence.

When the appointment has been made, it is a nice gesture to thank the client and mention that you look forward to being of service to her.

The principles of taking an appointment are the same whether the client telephones or visits the salon. Where the client attends the salon in person, however, remember that she can also make a *visual* assessment.

THE VISIT

It is no use creating this initial good impression if, on the client's visit, it is not followed through at every stage and by each member of staff with whom she makes contact.

THE RECEPTIONIST

This term should apply to *any* member of staff who makes initial contact with the client.

Immediately the client enters the salon, she must be welcomed in a friendly manner. After establishing the details of her appointment, it is the duty of the receptionist to assist her with the removal of her coat and to provide protective clothing.

If for any reason the client is to be kept waiting, an explanation should be given and she should be offered magazines and a cup of coffee or tea. Should the client light a cigarette, an ashtray should be made available without her needing to ask for one. This will be seen as genuine interest in her well-being.

The receptionist will, in most instances, be the last person to deal with the client before she departs, and will therefore be responsible for the final impression.

Apart from taking the money, the receptionist should remove any hairs from the client's clothing with a brush, ascertain whether another appointment is required, and finally assist the client with her coat before thanking her for her custom.

THE APPRENTICE

'The apprentice should be seen and not heard' was at one time a reality. Today however, the apprentice plays a very important role in the salon and is equally responsible for making the client's visit an enjoyable one.

Light, intelligent conversation will help to relax the client. Being ignored never makes anybody feel important or comfortable, neither does the apprentice who spends the entire shampoo period talking about her own personal business with another apprentice who is working alongside. Suggestions regarding special shampoos, conditioning treatments, etc., will be listened to appreciatively by the client, if the manner in which they are presented is polite and helpful.

THE STYLIST

A good relationship between the client and stylist is vital, and although this depends a great deal on the technical ability of the stylist to produce work of a high standard, this alone will not be sufficient to ensure a client's continued custom. In many instances, stylists who have less ability gain more regular business because of their personality and the interest they show in their clients.

A good stylist will greet his client as soon as possible and certainly before her hair is shampooed. Discussion to establish her requirements is vital, accompanied by suggestions regarding beneficial services and products.

Casual conversation should also be encouraged, particularly in areas of interest to the client.

It is also important that the stylist is a good listener and capable of recognizing when his client does not wish to talk but just relax and enjoy the treatment she is receiving.

SELLING SERVICES AND PRODUCTS

Many hairdressers believe that they are caring for their clients by not selling extra services or products to them, consequently presenting them with a smaller bill. This is not always the case.

It is true that hairdressers need to be salesmen to ensure profitability for the business and to pay their wages, but it should not be overlooked that by selling products, an improved service can be offered to the client, giving a more successful end result which leads to client satisfaction. Carried out correctly, salesmanship is of great benefit to the client.

The client visits the salon for one main reason. She wants to look and feel better for her visit. Because of this, most clients welcome suggestions from their stylist regarding the products and services which the salon has to offer and which will be advantageous to her. A client may only book an appointment for a shampoo and set; the biggest mistake we can make is in thinking that this is all she is willing or prepared to receive.

On the other hand, over-selling is as bad for business as underselling, and it is vital that the hairdresser's own personal gain does not become the only motivating factor. Put yourself in the place of the client. Would you continually return to someone who pushed you into buying products and services from which you obtained little or no benefit?

17

From a psychological point of view

Hairdressing fulfils the physical role of improving a client's appearance; it also fulfils a psychological role.

Our appearance is the initial way in which we promote ourselves to others; people who are aware that they are projecting a favourable image not only feel better for it, but also have more self-confidence and self-assurance. As hairdressers, we need to fulfil two psychological roles: firstly, that of helping the client to achieve the personal image which she herself wishes to project and secondly, promoting our own image in such a way that we gain the confidence of our clients.

Most people say that they do not judge others on their appearance. In the main this is incorrect. In personal relationships, there is no basic difference in the characteristics of people with pleasant or unpleasant physical attributes; we are, however, drawn to the former. Second impressions may destroy our earlier views and it is not unusual to find that people who have less physical attraction often try harder to promote a pleasant personality than those that do. In other words, 'beauty is only skin deep'.

The appearance of a salon as well as its staff, plays a crucial role in the success of the business. Clients expect receptionists to be clean, tidy, well-mannered, helpful, efficient and to have a pleasing personality. These same attributes are also expected in every other staff member. It is a known fact that physically attractive people (particularly if they have all of the other attributes mentioned) generally reap more success. Good examples of this are shown in the world of films and advertising.

Virtually all of us can improve our appearance and thus gain more favourable relationships with others. The careful use of make-up by female staff members plays a vital role in creating a favourable impression with the client, as will a smart appearance. Pride in appearance is equally as important for male staff.

The correct mental attitude towards both the client and the job itself, will mean that greater respect and favour are gained.

Psychologists have established a number of facts which should be given consideration. If a man wears a tie, this has a pronounced effect in gaining confidence. Eye contact between two people opens up a strong line of communication, looking away from a person cuts that line. By maintaining eye contact during discussion, even indirectly through a mirror, more attention will be paid by the listener to what is being said. This is invaluable when discussing extra salon services or products. Another fact is that the appearance of female staff is more often criticised by both men and women than that of male staff. This however, should not be used as an excuse by male staff.

LOOKING AT OURSELVES

We all have certain qualities which make up our personality. Our personality is our character. Because we sometimes feel that our 'true' character — known only to ourselves — is not always acceptable in every situation, we tend to adapt our behaviour and temperament to suit different occasions (or in the case of a hairdresser to suit different clients).

Most of us have ambitions of one kind or another, for example, ambitions to be successful in a job or to achieve a more attractive appearance These ambitions can only be achieved by our own efforts.

We communicate to others non-verbally as well as verbally, by such means as appearance, manner, body movements, etc., and hairdressing is responsible for helping others to create their non-verbal image — a very important task. Insincerity can be portrayed non-verbally by bodily gestures such as facial expressions, avoidance of eye contact, shrugging of shoulders, etc., often unconsciously, and so we need to put up 'a front' when dealing with a client. This should not be done with deception or insincerity.

Conclusion

As we have seen psychology plays a most important part in hairdressing, in the way we promote ourselves and our business to others. It is necessary for the hairdresser to be totally aware of everything which is relevant not only to his own personal success, but to that of the business, his clients and fellow staff members. In fact, his own success is determined by that of all other parties involved. Do not be afraid to carry out a self-analysis from time to time. Develop favourable points but be honest enough to admit bad points, particularly those referred to by others with whom you have regular contact. Take a personal pride in your appearance and work and remember that the most important person in the salon is the client.

Success is not handed out on a plate, it comes from having the right mental approach to everything.

ACCEPTING OUR RESPONSIBILITIES

Hairdressers have both a legal and moral obligation towards their client's safety and well-being. In any treatment where chemicals are used, there is always the risk of personal injury and damage to clothing and it is the hairdresser's duty to take every precaution to prevent this situation arising.

Neglect of this duty could lead to legal action being taken by a client to gain compensation for any distress which she has been caused. Such situations are seldom pleasant and can lead to bad publicity and consequently a loss of business.

Hairdressing salons are required to carry Public Liability Insurance to compensate for injury and damage caused on the premises, and in the event of a complaint being made, the following procedure should be followed:

1 Call the manager immediately.

2 Where a client has received a burn or cut, administer first aid.

3 If you feel that the client has in any way contributed to the injury or damage caused, make this known to the person in charge.

4 Do not admit liability or give the impression of admitting liability to a client.

5 Extend sympathy without being emotional.

6 Make written notes of all the facts and notify the insurance company immediately. This does not mean that you are admitting guilt.

From this point on, any correspondence from the complainant or her solicitor should be forwarded to the insurance company and not dealt with on a personal level.

Although there are exceptions to the rule, most damage or injury to a client is caused by not following correct procedures. Pay special attention to these points:

1 Always carry out a pre-disposition test where advised by the product manufacturer.

2 Never use a product in any other way than stated in the instructions. Pay special attention to quoted peroxide strengths when tinting or bleaching.

3 Ensure that adequate protection is provided to prevent chemicals from coming into contact with a client's clothing. Any chemicals which come into contact with the skin or eyes should be washed away immediately.

4 Before starting a perming, colouring, or bleaching operation, ensure that the client has no history of allergies to such preparations and check for skin abrasions and open wounds.

5 Do not be talked into carrying out any service where you consider there is a risk of damage to the hair or scalp.

6 Take care to place such items of equipment as heated rollers and curling irons out of easy reach of clients.

7 Ensure that all areas, particularly stairways, are well lit and uncluttered.

8 Keep record cards for every client having a bleach, colour or permanent wave and ensure that all the information it contains is accurate. Not only are record cards helpful to other members of staff during the absence of a particular stylist, they are a useful means of establishing that correct procedures were followed should a complaint arise.

A little common sense can save a great deal of distress.

Special note
Public Liability Insurance does not cover all types of injury which may occur in the salon. Suitable advice should be sought to ensure that adequate cover is effected.

18
Test your knowledge

The following selection of questions refer to the various chapters contained within this book.

A THE HAIR

1 What is hair?

2 What is the chemical composition of hair?

3 On average, a hair will grow for how long?

4 List the stages of hair growth in their correct order.

5 Hair is hygroscopic. What does this mean?

6 Give a brief description of the cuticle, cortex and medulla.

7 Food which is high in protein is good for the hair. Give four examples of high-protein foods.

8 Describe: a) the follicle b) the sebaceous gland c) the arrector pili muscle.

9 On which areas of the body is hair not found?

10 Draw a simple, labelled diagram of a skin section.

B HAIR AND SCALP DISORDERS AND DISEASES

1 What is the function of the acid mantle which covers the body?

2 Give three causes of poor hair condition.

3 Briefly explain pityriasis simplex and pityriasis steatoids.

4 What is seborrhoea oleosa?

5 How is canities brought about?

6 The term 'alopecia' is used to describe what condition?

7 What is pediculosis capitis, and how would you recognise it?

8 How would you detect hair breakage as opposed to natural hair fall?

9 What is dermatitis?

10 How can both hairdresser and client be protected from contracting dermatitis?

C HAIRDRESSING-RELATED CHEMISTRY

1 Define the following: a) elements b) compounds.

2 How is a salt produced?

3 What do you understand by the term 'emulsion'?

4 From what sources are the following oils obtained: a) vegetable b) mineral c) essential?

5 Name two solvents of resins and varnishes.

6 How is a sulphonated oil produced?

7 Why should any preparation which contains ammonia not be poured into a metallic bowl?

8 The following are symbols for which chemicals: a) H b) K c) Fe d) Cl e) Mg f) O?

9 Describe how the strength of hydrogen peroxide is tested.

10 Draw a diagram of the pH scale and show the approximate positions on the scale of: a) the hair b) low pH perms c) high pH perms d) tints and bleaches e) hair straighteners.

D ELECTRICITY

1 What is electricity?

2 Two types of electric current can be generated. Name and briefly describe them.

3 A hairdryer contains an electrical element. What is the purpose of this element?

4 Why are fuses inserted into an electrical circuit and into the plugs attached to electrical equipment?

5 What initial course of action would you take in the event of someone receiving an electric shock?

6 Briefly define the following electrical terms: a) volt b) ampere c) watt d) ohm.

7 What do you understand by the terms 'insulators' and 'conductors'? Give two examples of each.

8 How is static electricity created, and what problems can it cause when styling the hair?

9 How is electricity produced in a torch battery?

10 Secondary batteries are capable of storing electricity fed into them from a mains supply. What is their big disadvantage?

E SCALP MASSAGE

1 What is the purpose of scalp massage?

2 Explain the movements used in effleurage massage, and petrissage massage.

3 Describe the vibro massager and state its function.

4 What precautions must be taken before starting a high–frequency treatment?

5 How is a high-frequency treatment carried out using the condenser electrode?

F SALON HYGIENE

1 List all the factors which you consider important in maintaining a clean, tidy, and hygienic salon.

2 What are the best methods for adequately sterilizing the following: a) glass and porcelain objects b) metal instruments c) towels, gowns and capes?

3 Describe the vaporization method of sterilization.

4 What are the advantages of the ultraviolet method of sterilization over the vaporization method?

5 Write a short essay on the importance of personal hygiene in the salon, including all the factors which need to be considered.

G SHAMPOOING AND CONDITIONING

1 What is the function of shampoo?

2 What disadvantages do shampoos containing soap have?

3 How are soapless shampoos manufactured?

4 List the benefits expected from a good shampoo.

5 Give examples of usage for the following types of shampoo: a) base b) cream c) oil d) lemon e) medicated.

6 Describe, as fully as you can, the correct shampooing procedure.

7 What benefits should be obtained from the use of a surface conditioner?

8 Explain the correct procedure for applying a surface conditioner.

9 What advice could you give to a client suffering from a greasy scalp condition, other than recommending a course of specialized treatments?

10 A head of hair is very dry and chemically damaged, but the scalp is greasy. How would you tackle this problem?

11 What do you understand by the term 'hair re-structurant'?

12 When would you recommend the use of a re-structurant as opposed to a surface conditioner?

H PERMANENT WAVING/ STRAIGHTENING

1 Explain, in simple terms, how a cold wave works.

2 What are the main factors to be considered when choosing a permanent wave lotion?

3 What is meant by the term 'directional wind'?

4 Other than cold waving, what other methods of permanent waving do you know of?

5 How do style supports differ from permanent waves?

6 What chemical action occurs during the neutralizing stage?

7 When and why would you carry out a compatibility test, before proceeding with a permanent wave?

8 What may be the result if a perm curler rubber is twisted, or pressing on the hair, during processing?

9 Give three possible reasons for there being no result on completion of a permanent wave.

10 What damage can be caused to the hair, by using a lotion which is too strong for the hair type, or by over-processing?

11 When may it be necessary or desirable to straighten a head of hair?

12 Explain the method employed when straightening a naturally curly head of hair.

I COLOURING AND BLEACHING

1 What are: a) the primary colours of pigment b) the secondary colours of pigment?

2 Explain what happens to colour when subjected to the following light sources: a) daylight b) tungsten light c) fluorescent light.

3 When and why is a pre-disposition test (skin test) carried out?

4 Give a brief description of each of the following: a) temporary colours b) semi-permanent colours c) permanent tints.

5 List as many points as you can which need to be considered when carrying out a head assessment.

6 How would you make an application of permanent tint to lighten virgin hair?

7 In what circumstances may it be necessary to use a colour stripper?

8 Describe the procedure for stripping colour from the hair.

9 When bleaching hair, why is it seen to lighten through varying stages of red and yellow?

10 Oil bleaches, emulsion bleaches and powder bleaches are each particularly suited to certain bleaching requirements. State when each might be used.

11 Explain how a bleach application is made to virgin hair.

J HAIR STYLING

1 List the considerations you feel need to be taken into account to ensure the suitability of a style for each individual client.

2 What do you understand by the following cutting terms: a) layered cut b) graduated cut c) reverse graduation?

3 What advantages do you feel are obtained by blow-drying many of the more modern styles, as opposed to roller setting?

4 How can the maximum amount of root lift be achieved, when using a brush and blow-dryer?

5 When setting hair, why should the rollers not be positioned in straight rows?

6 Explain how hair is able to take on a new shape when set around a roller.

7 How are rollers positioned in relation to their section for: a) maximum lift b) when the hair should lie closer to the head?

8 What is reverse curling and how is it achieved?

9 Backbrushing and backcombing both perform a similar purpose. What is this purpose?

10 Why does a setting lotion prolong the life of a set?

11 Explain the correct procedure for applying a setting lotion.

12 What do you understand by the term 'dual-purpose setting lotions'?

K ASPECTS OF BARBERING

1 Using diagrams, explain the differences between a hollow-ground razor and a French or solid razor.

2 Name the two types of strop and explain their purposes.

3 What is meant by the following terms: a) first time overshave b) second time overshave?

4 Draw a diagram of a razor, listing the parts.

5 Describe how a hollow-ground razor is set.

6 In what instances are hot towel treatments carried out?

7 Name and describe the three movements used when carrying out a facial massage using the hands.

8 Why, and how, are friction lotions applied to the scalp?

9 Explain the procedure employed when trimming a beard and moustache.

10 Describe how a facial massage is given using the vibro massager.

Terms commonly used in hairdressing

Acid Mantle	The natural acid covering of the body.
Allergy	A condition of abnormal sensitivity to an irritant. It may occur at any time.
Alopecia	A medical term for the loss of hair; baldness.
Amino Acids	Nitrogenous organic compounds. They constitute the essential parts of a protein molecule.
Ammonia	A colourless gas. It dissolves rapidly in water to form ammonium hydroxide, a strong alkali.
Anagen	The active growing stage of a hair bulb.
Antioxidants	Any substance which delays deterioration by oxidation.
Antiseptic	A solution which prevents or inhibits the growth of micro-organisms.
Astringents	Substances which tighten the skin such as witch hazel.
Atom	The smallest particle of an element. Atoms combine with similar particles of other elements to form compounds.
Bacteria	Minute living organisms, which often cause disease.
Balance	A term used by barbers to indicate the comparative weight of a razor blade to that of its handle.
Base	The combination of a metallic element with oxygen. When dissolved in water it is called an alkali.

Biodegradable	Any substance or material which is easily broken down by mico-organisms for use as nourishment.
Canities	Medical term for hair which has lost its colour pigment, causing it to look white.
Carbon Tetrachloride	A solvent sometimes used in dry-cleaning processes. It is highly toxic.
Catagen	The stage of hair growth where the bulb produces less and less new cells until ceasing completely.
Cells	(Found in the living body.) Extremely small, separate masses of protoplasm, surrounded by membranes. There are around one million million cells in the human body.
Chignon	A hairpiece attached to the head to achieve a particular style or add bulk.
Colour Stripping	The process of removing artificial colourants from the hair.
Cortex	The central section of the hair where the colour pigment is contained.
Cuticle	The hard, protective outer layer of the hair shaft.
Cysteine	Cystine which has been reduced by a reducing agent such as ammonium thioglycollate in permanent waving solutions
Cystine	A sulphur-containing amino acid. It is produced in the digestion of proteins.
Depilatory	A substance which removes hair.
Depth	A term used to describe how dark or light a colour is.
Dermatitis	An inflammation of the skin, which can be severe in certain cases.
Dermatology	The study of the skin covering its diseases and treatment.
Disinfectant	A solution which destroys bacteria.
Distillation	The purification of liquids through evaporation followed by condensation.
Dressing	A term used to denote both a particular hairstyle and a hair application such as brilliantine.

Effleurage	A method of massage using the open palms of the hands.
Electrology or Epilation	Destruction of the hair carried out by inserting a fine needle into the hair follicle alongside the hair and passing an electric current through it.
Element	Any substance which cannot be broken down by ordinary chemical methods into simpler substances.
Emulsion	A substance made up of two immiscible liquids, such as oil and water. One liquid is made to disperse through the other by means of an emulsifying agent.
Eruption	A term often used to describe the outbreak of skin blemishes, such as acne.
Evaporation	The process of changing from a liquid form to a vapour form. The speed of evaporation is dependent on temperature, being very rapid at boiling point.
Exothermic	A term used to describe a chemical reaction in which heat is created.
Fibrous	Made of fibres.
Germs	Microscopic organisms which cause diseases — bacteria and viruses.
Gland	An organ which produces a particular chemical substance which it then releases; for example, the sebaceous gland produces sebum which it releases into the hair follicle.
Grind	A term used by barbers to describe the shape of a razor blade when viewed in cross-section.
Humidity	The water content present in the atmosphere.
Hydrogen Peroxide	A colourless liquid used in bleaching products.
Hygroscopic	The ability to absorb moisture.
Infestation	Attacks on the body by small parasitic animals, for example, head lice.
Infra-red	Invisible heat rays found in sunlight. Created artificially by medical science for the treatment of certain ailments.
Ion	An atom with a positive or negative charge. Found in solutions and some solids and salts.

Keratin	An insoluble protein of which nails and hair are composed.
Lanugo Hair	Fine, downy hair, found on the face, arms, etc. It is short hair which is hardly noticeable and has no medulla. It is often found on the human foetus before birth.
Lipids	Fatty substances which are insoluble in water but soluble in alcohol.
Malleable Block	A false head of strong canvas stuffed with chopped cork. Used for the dressing out of wigs.
Matter	Any substance occupying space and having weight.
Medulla	The core which runs through the middle of the hair shaft.
Melanin	Colour pigment produced by cells called melanocytes and found in the hair and skin.
Molecule	A group of two or more atoms, for example, water (H_2O) molecules are composed of two atoms of hydrogen plus one of oxygen.
Neutralization	The term used to describe the process by which a salt is formed by an acid/alkali mixture.
Organism	Any individual living unit such as a plant, bacteria, animal, virus.
Over-processing	A term used to describe when lotion is left on the hair for too long.
Oxidation	The reaction of a substance with oxygen where there is a loss of electrons from one chemical type to another.
pH	A measure which is used to express the hydrogen ion content of a solution, indicating varying degrees of acidity or alkalinity.
Pediculosis Capitis	The infestation of the scalp by head lice.
Peroxometer	Instrument used to measure the strength of hydrogen peroxide.
Petrissage	A method of massage using the fingertips.
Pin Curl	A small section of hair wound in a circle and secured with a hairpin to make it curl.
Pityriasis	A medical term used to describe scurf or dry dandruff.

Polypeptides	Molecules composed of a number of amino acids linked together by peptide linkages.
Polyvinylpyr-rolidone (PVP)	A water-soluble, synthetic resin, used in the manufacture of hair sprays and setting lotions.
Porosity	A readiness to absorb moisture. Damaged hair will have a greater degree of porosity than normal healthy hair.
Porous	A substance is said to be porous when it is full of small holes and has an ability to absorb moisture.
Pre-disposition Test	A skin test carried out to ensure the skin does not show any allergic reaction to certain chemicals.
Psoriasis	A skin condition characterized by dry, hard, silvery scales with a reddish border.
Public Liability Insurance	An insurance policy which must be carried by all hairdressing salons, which compensates for any injury or damage caused on the premises.
Pull Burn	A painful scalp condition due to excessive tension on the hair during permanent waving, accompanied by lotion entering the follicle.
Reagents	All substances used for detecting, examining or measuring other substances.
Reducing Agent	A chemical which removes oxygen from, or adds hydrogen to a compound.
Resorcinol	A colourless liquid used in making dyes.
Re-structurant	A special conditioning product which penetrates the hair where any cystine linkages are weakened.
Salts	Substances formed by combining an acid with an alkali.
Seborrhoea	A term to describe acute dandruff.
Sebum	The oil secreted on to the skin by the sebaceous glands.
Soluble	Capable of being dissolved.
Solvent	Substances into which other substances will dissolve.

Sterile	Totally germ free.
Sulphonated Oils	Oils made to react chemically with concentrated sulphuric acid. Used in the manufacture of soapless shampoos.
Surface Conditioner	A conditioning product designed to lubricate the hair shaft, leaving a fine film which reflects the light.
Surfactants	Detergents used for the cleaning of many structures including hair.
Tapotement	A method of massage using the fingertips with a light tapping action.
Telogen	The resting stage of a hair bulb during a growth cycle.
Temper	A term used by barbers to describe the hardness of a steel blade of a razor.
Terminal	A medical term given to the coarse hair found on the head, armpits, pubic areas, beard and moustache.
Tone	A term used to describe the actual colour of hair (gold, warm, ash etc.).
Toupee	Small wig made for men to conceal the type of baldness which is generally at the crown of the head and forehead.
Trichologist	A qualified person who practises as a consultant for hair and scalp disorders.
Trichology	The science of hair; its research and investigation.
Triethanolamine Lauryl Sulphate	An organic salt used in the manufacture of soapless shampoos.
Ultraviolet (UV)	Strong, invisible light rays, produced by the sun and essential to life. Ultraviolet rays can be artificially produced by special lamps, but should only be used under expert guidance.
Vellus	A medical term given to the soft hair which covers most areas of the body.
Virgin Hair	Hair which has not undergone any form of chemical treatment.

Index